ST(P)
Technology
Today Series

Engineering Drawing for TEC Level II

ST(P) TECHNOLOGY TODAY SERIES

A full list of titles in the series is available from the publishers on request, post free.

The series covers mathematics and statistics for Mechanical and Electrical Engineering; Physical, Applied and Mechanical Engineering Science; Engineering Instrumentation and Control; Engineering Design; Electronics; Building Construction.

Engineering Drawing for TEC Level II

D. OLDHAM C.Eng., M.I.Mech.E., Cert.Ed.

Rotherham College of Arts and Technology

Stanley Thornes (Publishers) Ltd

First published in 1982 by Stanley Thornes (Publishers) Ltd, Educa House, Old Station Drive, Leckhampton Road, Cheltenham GL53 0DN.

British Library Cataloguing in Publication Data

Oldham, D.
Engineering drawing for TEC level II.
1. Engineering design
I. Title
620'.00425 TA174

ISBN 0-85950-387-9

Typeset by TECH-SET, 3 Brewery Lane, Felling, Gateshead, Tyne & Wear.
Printed in Great Britain by Ebenezer Baylis & Son Ltd,
The Trinity Press, Worcester, and London.

to Maureen, Simon and Mark

Contents

Preface

This book is intended for students studying for the TEC Certificate and Diploma in Mechanical and Production Engineering (TEC programme A5).

It covers the general objectives, with liberal illustrations, for Engineering Drawing II, by introducing loci, mechanisms, cams and bearings and furthering the student's knowledge of screw threads and their applications.

It informs the student that components, in addition to errors of size, can have errors due to form, attitude and location, and introduces the student to geometrical tolerancing to BS 308: Part 3: 1972 (ISO).

The student is instructed in the method of producing working engineering drawings, to British Standard Specifications, by the use of a case study to give a formative approach.

A special feature of the book is the self-tests, set at the end of each chapter, to test the student's knowledge of the topic area. Long-answer exercises are also given at the end of each chapter to test this knowledge to a deeper extent.

Also in this series: *Engineering Design for TEC Level III*

An additional book is planned which covers Engineering Drawing I.

Sheffield 1982 D. OLDHAM

Acknowledgements

I wish to express my thanks to Mike Brooks, for his helpful comments on the manuscript, and to Derrick Redford, who checked the manuscript in detail.

I am grateful for the kind permission, granted by the British Standards Institution, to use material based on information contained in their publications, particularly BS 308: 1972. Copies of the complete British Standards may be obtained from the Institution.

My thanks are also due to the following companies for their help in providing information and drawings:

RHP Bearings Ltd
Boneham & Turner Ltd
The Hymatic Engineering Co. Ltd
Record Ridgway Tools Ltd.

I would also like to express my appreciation to the publisher, for the friendly co-operation and helpful advice, to Mrs Pat Smith, for the excellent typing of the manuscript, and to the College Librarian, Miss B.C. Lim, ALA, and her staff, for their help.

Finally, my thanks to my wife, Maureen, for her help in checking the manuscript and for her encouragement during its preparation.

Note: Dimensions on all drawings are expressed in millimetres (mm) unless otherwise stated.

SECTION A

LOCI AND MECHANISMS

1 The Construction of Loci

After reading this chapter you should be able to:

- ★ appreciate the importance of constructing loci in relation to mechanism operation (G);
- ★ define the locus of a point (S);
- ★ relate the locus of a point to a circle, ellipse and cycloid (S);
- ★ explain the importance of determining the locus of a point on a moving mechanism (S);
- ★ explain the principle of quick return motion (S);
- ★ construct loci of points for given mechanisms, for example, slider–crank, four-bar chain (S).

(G) = general TEC objective
(S) = specific TEC objective

LOCUS OF A POINT

The locus of a point is the path on which it moves. (The plural of locus is 'loci'.) A line can be considered to be the successive positions of a moving point. A locus is dependent on the motion of a point and is found by joining a number of positions which are made by a point as it moves.

Fig. 1.1 shows a football in flight. As it moves it takes up successive positions 1, 2, 3, etc. Joining these positions will give the locus of the ball.

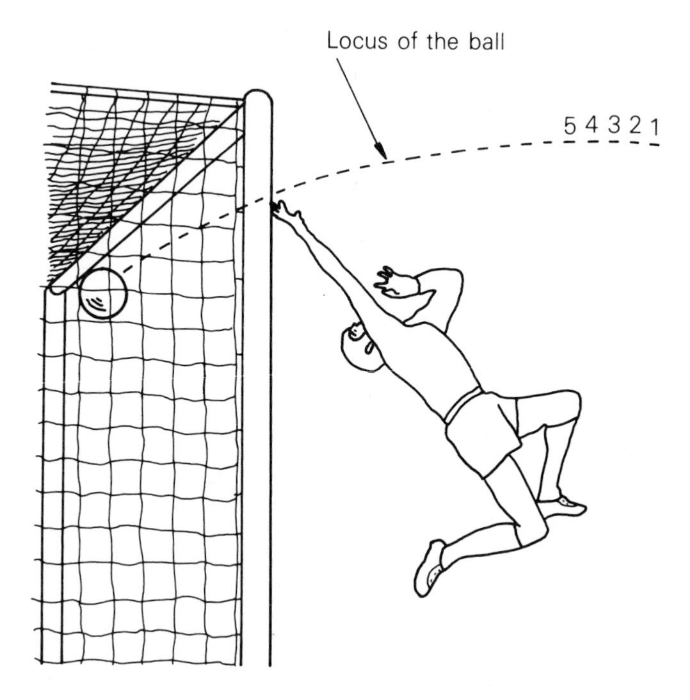

Fig. 1.1 Locus of a point.

If a point moves at a constant distance from a straight line, the locus of the point will be a straight line locus parallel to the original line — see Fig. 1.2.

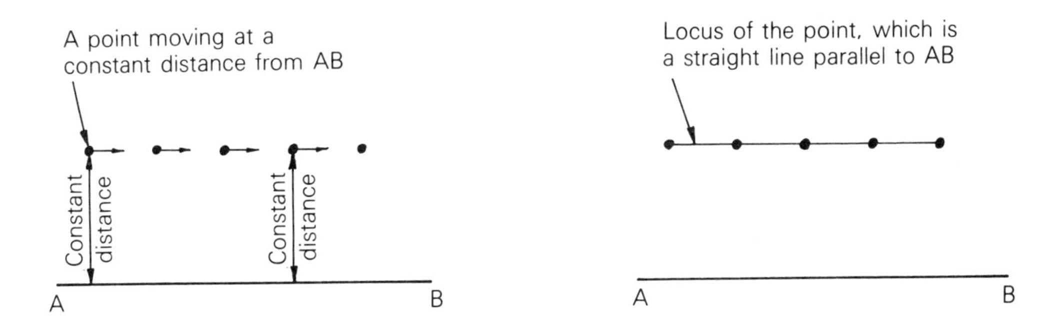

Fig. 1.2 Locus of a point moving at a constant distance from a straight line.

If a point moves at a fixed distance from another fixed point, the resulting locus will be an arc of a circle, or the complete circumference of a circle — see Fig. 1.3.

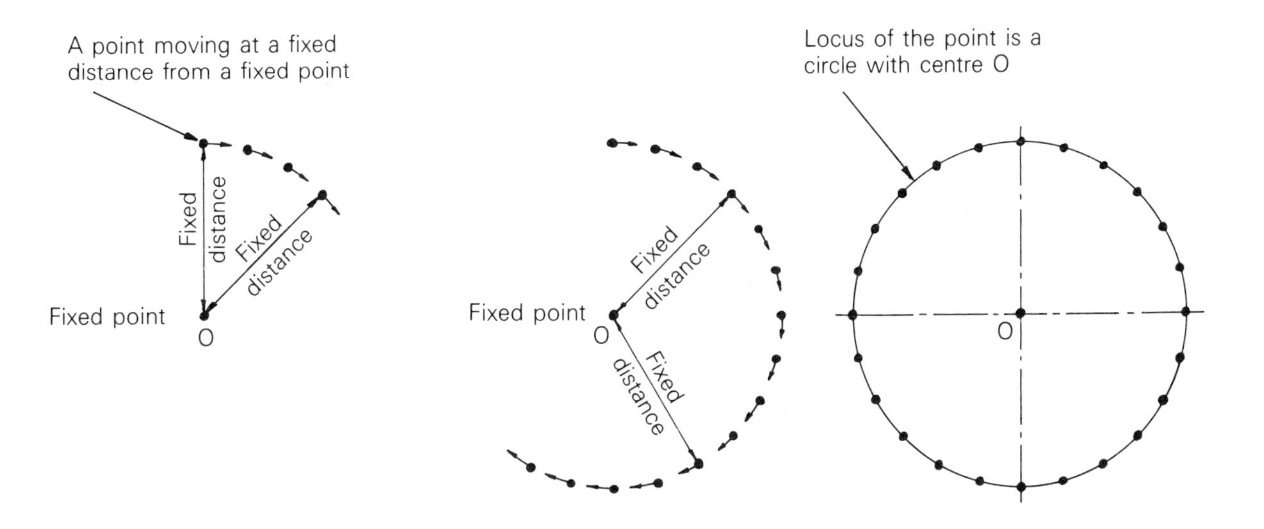

Fig. 1.3 Locus of a point moving at a fixed distance from a fixed point.

Fig. 1.4 shows the locus of a point which moves so that it is always equidistant from two fixed points X and Y. To obtain this locus, compasses should be set at a distance greater than XY/2 and, using X and Y as alternate centres, arcs should be struck either side of the line XY. The intersections of these arcs will produce a number of points which, when joined, will make the locus. It can be seen that the locus is the bisector of the line XY.

Fig. 1.5 shows a circle, centre O, resting on a horizontal line AB. If the circle is rotated through one-third of a revolution, the centre will have moved to position O_1. If the circle is rotated through two-thirds of a revolution, the centre will have moved to position O_2. Finally, if the circle is rotated through one complete revolution, the centre will have moved to position O_3. The locus of the centre, a straight line, can be obtained by joining the points O, O_1, O_2 and O_3, and it can be seen to be parallel to the horizontal line AB.

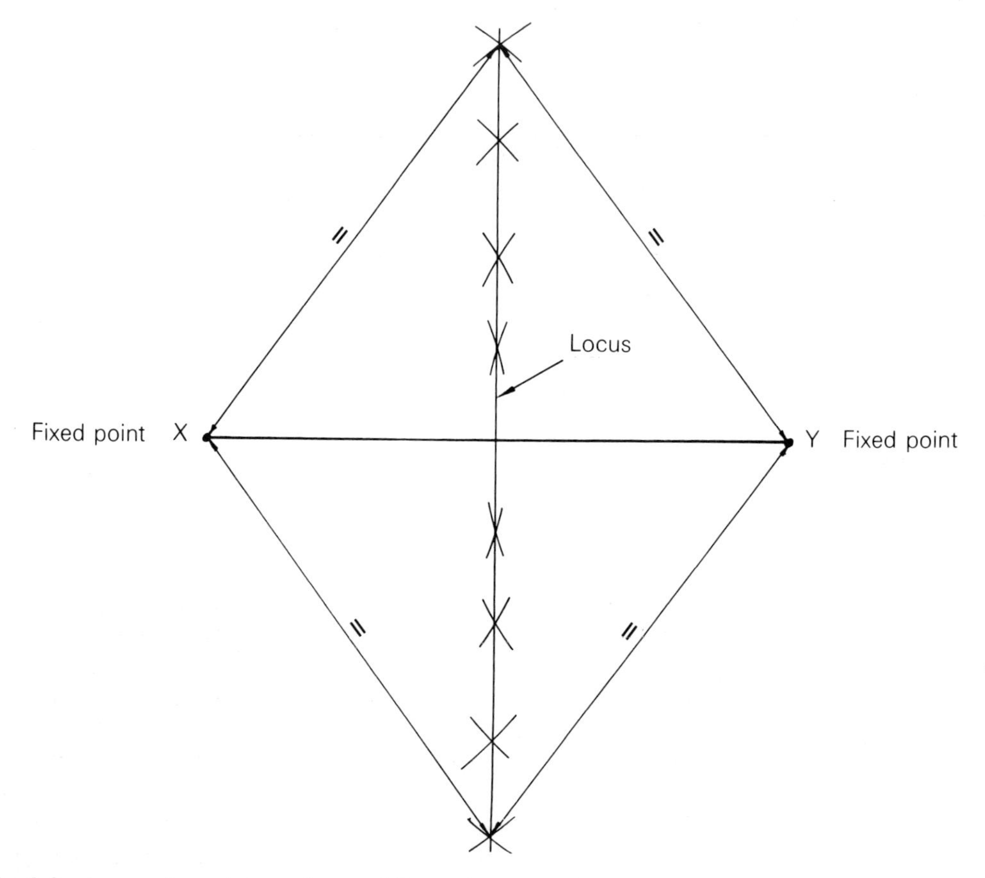

Fig. 1.4 Locus of a point moving equidistant from two fixed points.

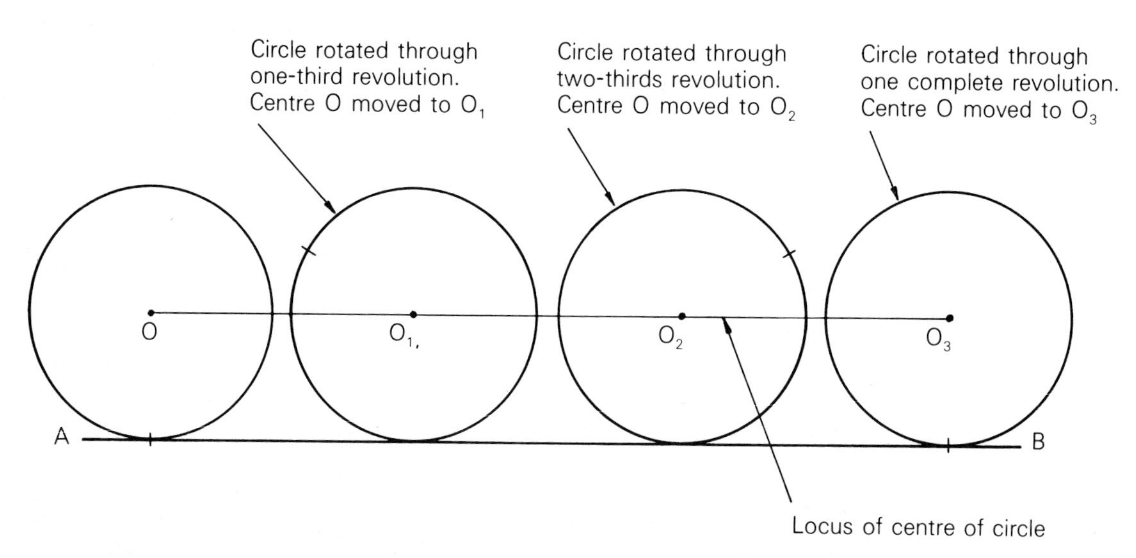

Fig. 1.5 Locus of the centre of a circle.

Cycloid

When a circle rolls along a line, a point on its circumference produces a locus, which is called a cycloid, as shown in Fig. 1.6.

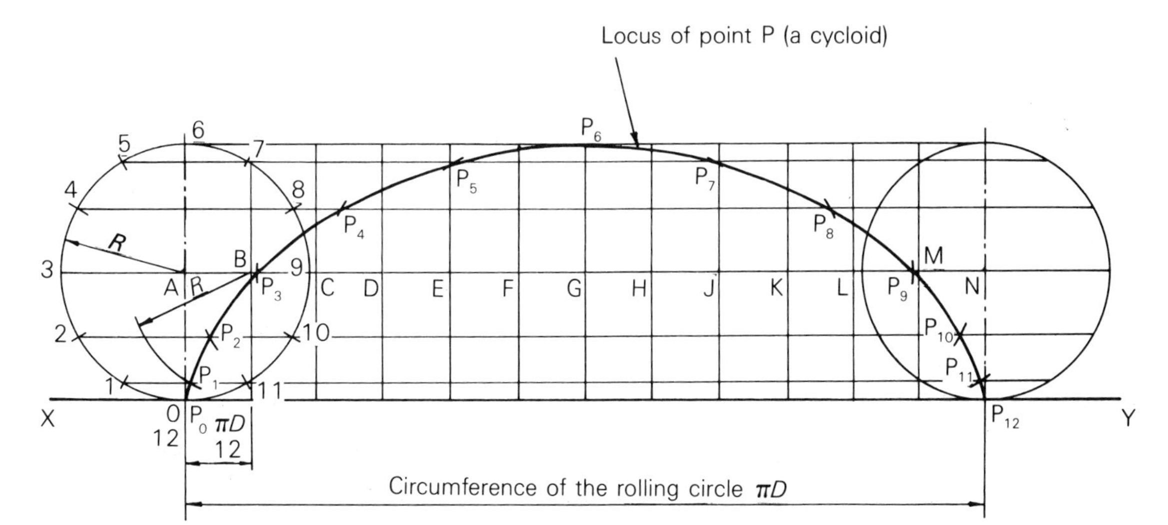

Fig. 1.6 Cycloid.

Consider a point P on a circle, of radius R, which rolls along line XY. In one complete revolution the circle will have travelled a distance equal to its circumference, i.e. $2\pi R$ or πD. To plot the locus divide this distance into, say, 12 equal parts marking their positions, in line with the centre of the circle, A, B, C, etc. Divide the circumference of the circle into 12 equal parts 0, 1, 2, 3, etc., and construct horizontal lines through each position. As the circle rolls one-twelfth of a revolution, the centre A will move to position B and the point P, initially at position P_0, will move to P_1. The position P_1 is found by striking an arc, of radius R, from centre B to the position where it crosses the horizontal line constructed from circle position 1.

As the circle rolls a further one-twelfth of a revolution, its centre will move to position C and the point P will move to P_2, which is similarly found by striking an arc, of radius R, from centre C to the position where it crosses the horizontal line constructed from circle position 2. Similarly positions P_3, P_4, P_5, etc., can be determined. The locus, a cycloid, can be drawn by joining up all these points.

Ellipse

An ellipse is the resulting shape of the locus of a point which moves so that the sum of the distances from two fixed points is constant. Each of these fixed points is called a focus (plural 'foci').

Fig. 1.7 shows how an ellipse is generated from the locus of a point. A and B are two fixed points and point P moves so that the sum of the distances from A and B is constant, i.e. $X + Y = X_1 + Y_1 = X_2 + Y_2$, and P moves successively to positions P_1, P_2, etc.

When point P has completed one revolution, an ellipse has been formed. The sum of the distances of P from the fixed points A and B equals the distance JK on the ellipse which is known as the *major axis* of the ellipse. LM is the *minor axis* of the ellipse and is determined by the positions where AL = AM = BL = BM.

In practice, the ellipse is a common occurrence so it is important to be able to construct it readily. Several methods are given for its construction, both as a true ellipse and as an approximate curve using arcs. In most cases the major and minor axes are known or can be easily determined.

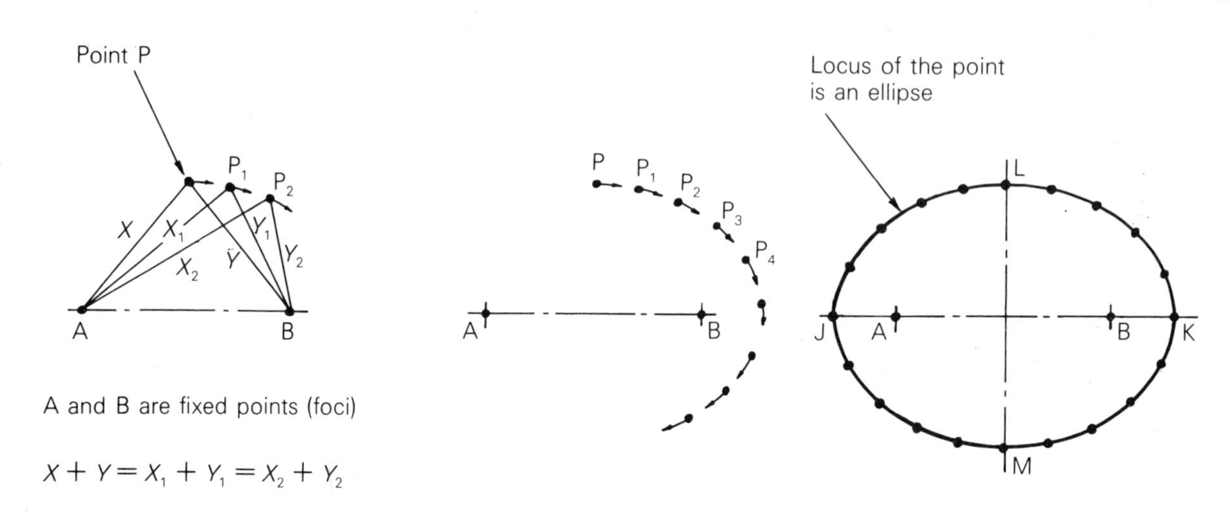

Fig. 1.7 Ellipse.

Construction for the ellipse

Method (a) Fig. 1.8a shows the trammel method of constructing an ellipse. The major and minor axes JK and LM are drawn. A trammel is made from paper or thin card and the distances XY, equal to half the major axis, and YZ, equal to half the minor axis, are marked on it. The trammel is moved keeping Z always on the major axis and X always on the minor axis. Point Y will trace out the path (the locus) of an ellipse.

Method (b) Fig. 1.8b shows the auxiliary circle method of ellipse construction. Two circles with centre O are drawn equal in diameter to the major and minor axes of the ellipse. The circles are divided into twelve equal parts (this is easily done using a 60°/30° set square) and vertical lines are drawn from the points where the radial lines cross the larger diameter circle, i.e. from points 1, 2, 3, etc., and horizontal lines are drawn from the points where the radial lines cross the smaller diameter circle, i.e. from points a, b, c, etc. The intersections of these horizontal and vertical lines give points through which the curve of the ellipse can be drawn.

Method (c) Fig. 1.8c shows the rectangle method of construction of an ellipse. A rectangle is constructed equal in length and breadth to the major and minor axes of the ellipse. KO and KN are divided into the same number of equal parts, numbering the points from K. Through these two sets of points lines are drawn respectively from M and L. Their intersections give points through which part of the curve of the ellipse can be drawn. Similar constructions in the other parts of the rectangle will give points through which the remainder of the ellipse can be drawn.

Method (d) In each method of ellipse construction shown above, the final curve is drawn through a number of plotted points. This usually involves the student in freehand work, resulting in some inaccuracy, although this can be reduced somewhat by the use of french curves or Flexicurves. An instrument construction method is shown, however, in Fig. 1.8d, where use is made of four arcs to obtain an approximate ellipse whose accuracy and construction depend on its dimensions. The major and minor axes JK and LM are drawn. Point X is positioned along the minor axis at a distance from centre O equal to the difference between the lengths of the major and minor axes, i.e. JK — LM. Point Y is positioned along the major axis at a distance from O equal to three-quarters OX. XY is joined and produced to Z, the tangent point for the two arcs. X is the centre for radius R_1 and Y is the centre for radius R_2. A similar construction is needed to obtain the centres for R_1 and R_2 in the other portion of the ellipse.

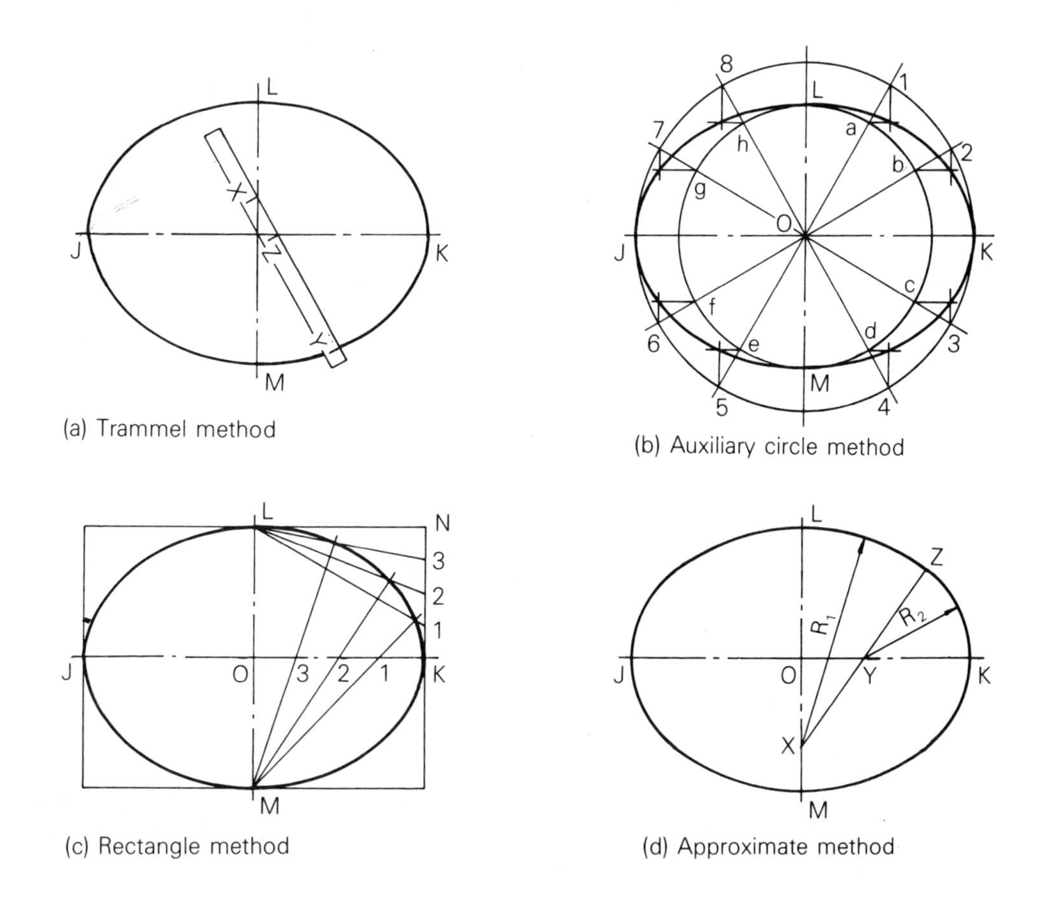

Fig. 1.8 Constructing an ellipse.

LOCUS OF A POINT ON A MOVING MECHANISM

Many situations arise in the design of mechanisms where the locus of a moving part must be determined. For example, during the action of a mechanism some element may need clearance to allow access or to stop it jamming a fixed part; some elements may need to be guarded to avoid the risk of accidents to operators, users or other personnel.

Fig. 1.9 shows an offset slider-crank mechanism (the slider-crank mechanism is discussed later). As the crank rotates it causes the connecting rod to oscillate to the extreme positions shown in Figs 1.9a and 1.9b. It is necessary to plot the locus of the moving part so that a suitably shaped guard may be fitted to avoid danger to personnel.

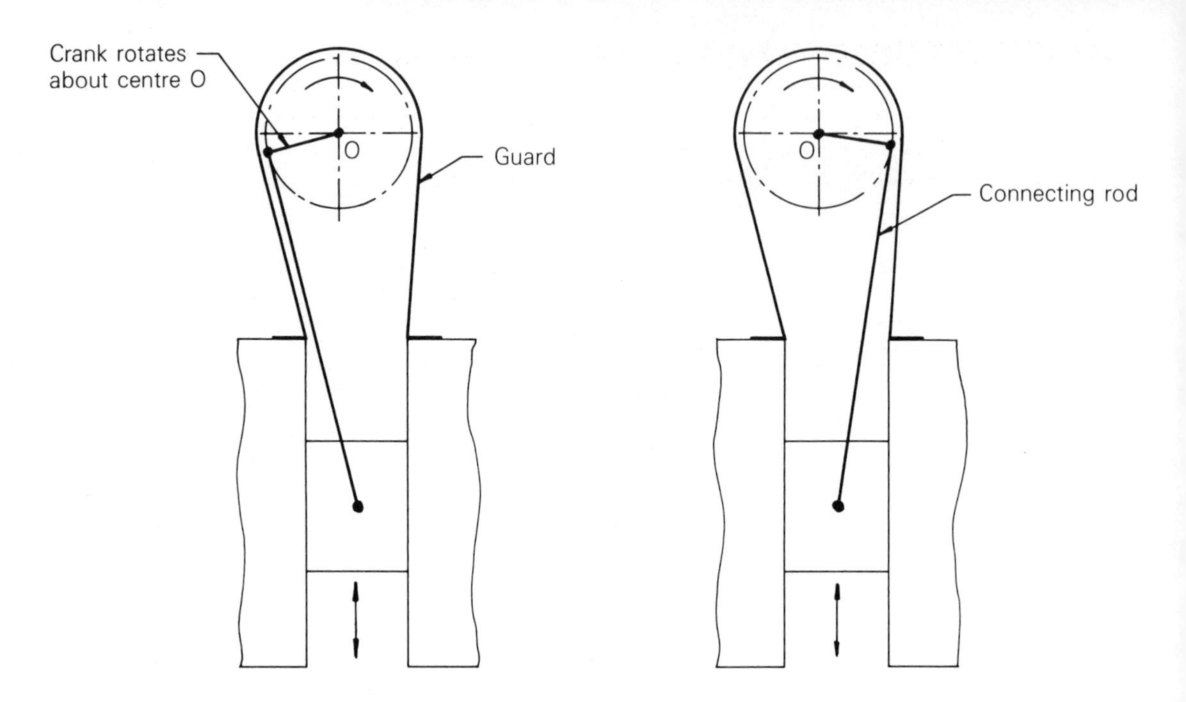

Fig. 1.9 Determining the locus of a moving part for guarding purposes.

Fig. 1.10 shows a conveyor system used to transport wire baskets during a manufacturing process. Conveyors of this type usually have two chains running parallel to each other and joined only by a hanger bar from which the component is suspended. It is necessary to determine the loci of the extremities of the basket, in this case the leading and trailing bottom corners, so that a guard can be installed which will completely shield its path.

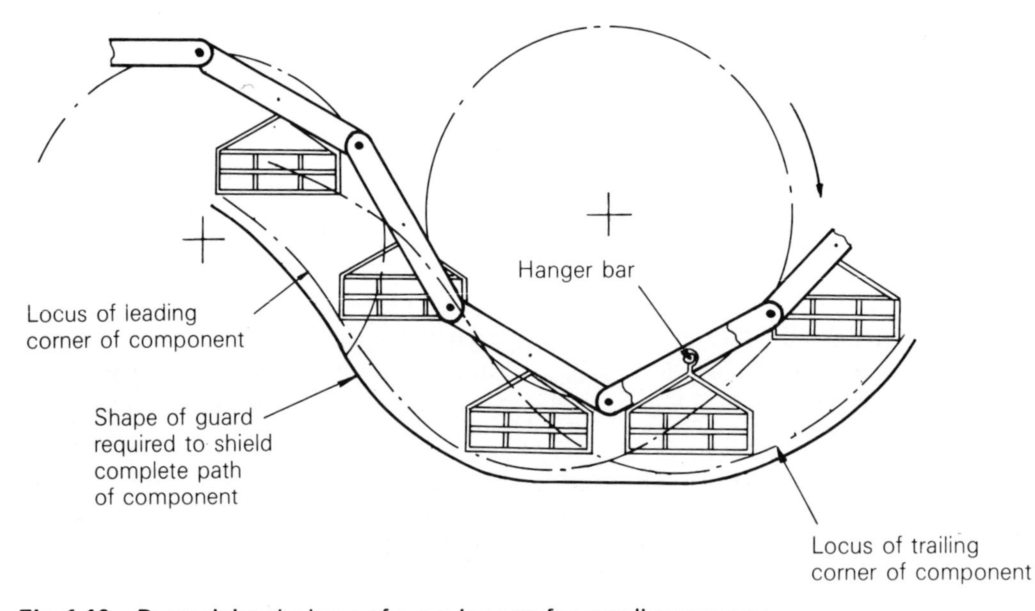

Fig. 1.10 Determining the locus of a moving part for guarding purposes.

QUICK RETURN MOTION

In straight-line cutting machine tools, such as shaping, planing and slotting machines, the cutting tool removes metal only during the forward stroke. The return stroke is an idle stroke and in the interests of production efficiency it should be completed in the quickest possible time.

One way in which quick return motion is achieved is by the method shown in Fig. 1.11. The system incorporates a crank wheel, rotating at a constant speed, on which is mounted a crank pin. Fig. 1.11a shows the mechanism at the beginning of the forward stroke with the crank pin at position A. As the crank wheel rotates the crank pin moves in a slot in the link L and causes it to pivot about point P. Fig. 1.11b shows the mechanism with the forward stroke underway. When the crank pin has reached position C, shown in Fig. 1.11c, the forward stroke is completed. As the crank continues to rotate, the crank pin moves to position D (Fig. 1.11d) and the return stroke is being effected with link L moving in the opposite direction.

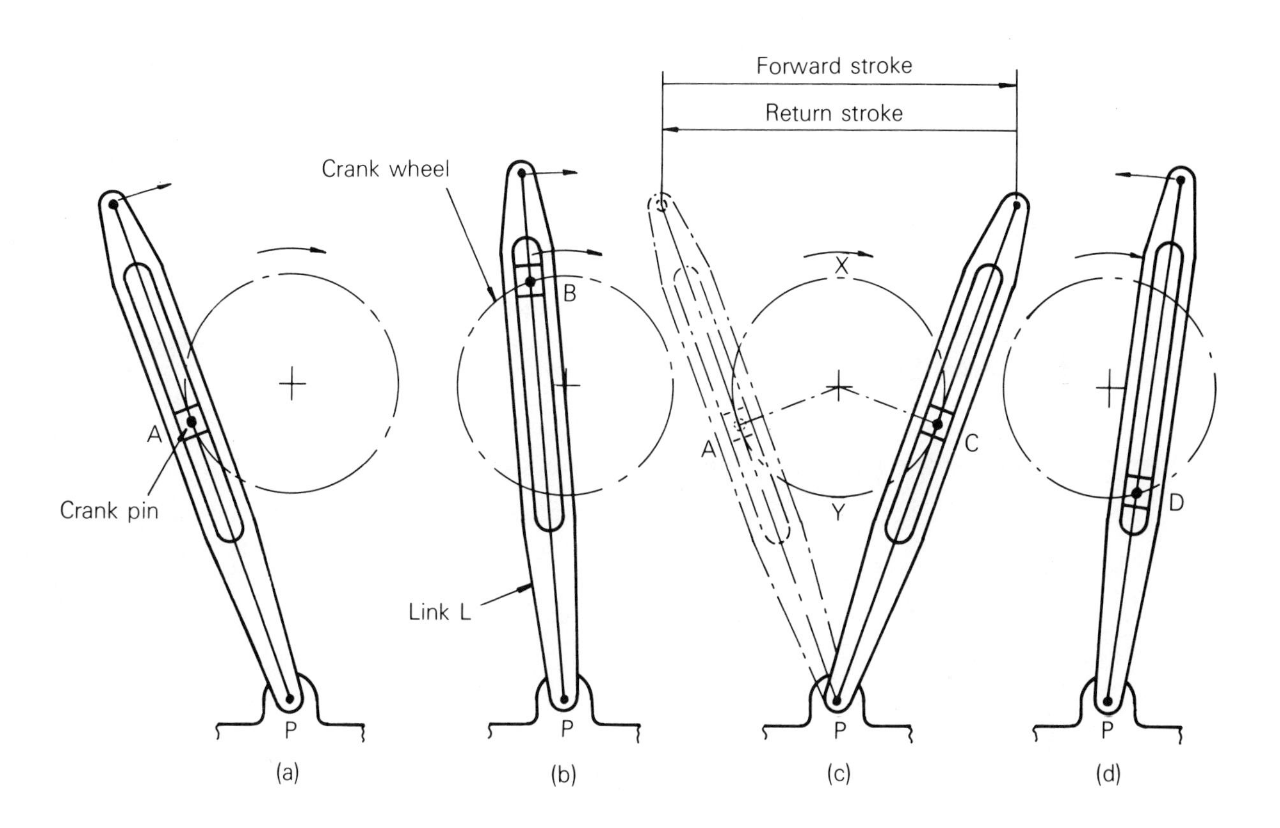

Fig. 1.11 Quick return motion.

During the forward stroke the crank pin moves from position A to C through the arc AXC, and the return stroke is made while the crank pin moves from C to A through the arc CYA. Because the crank rotates at a constant speed, the time taken for the crank pin to travel through arc CYA is much quicker than the time taken to travel through arc AXC. Consequently the return stroke is quicker than the forward stroke.

Slider–Crank Mechanism

A simplified slider-crank mechanism, shown in Fig. 1.12, is a very common approach adopted to convert rotary motion to reciprocating motion as in a power hacksaw. The converse of this principle is used to obtain rotary motion from reciprocating motion in an internal combustion engine.

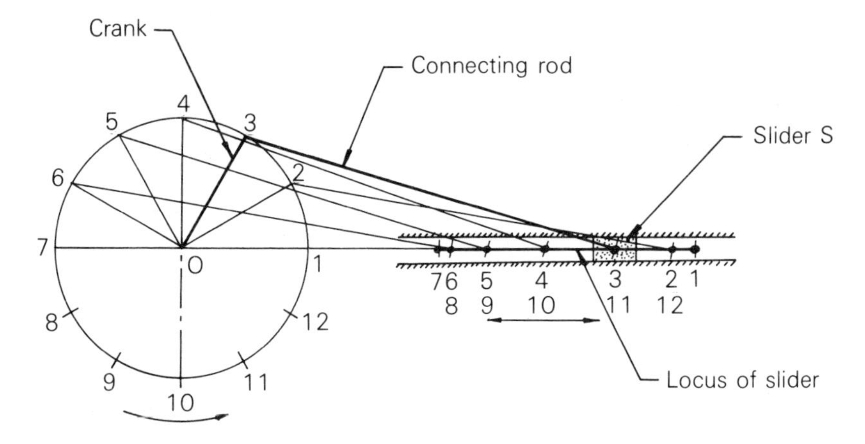

Fig. 1.12 Locus of the slider in a slider-crank mechanism.

The crank rotates about centre O and the connecting rod, which is attached to it, causes slider S to move. The slider is constrained, by guides, to move in a straight line only. As the crank rotates from position O1 to O2, O3, O4, etc., the slider moves from position 1 to 2, 3, 4, etc., until it reaches the limit of its travel at 7. At this stage the crank has rotated to position O7. As the crank moves to positions O8, O9, etc., the slider motion is reversed until it arrives back at its starting position 1 as the crank returns to O1. The straight-line path 1 to 7, and its return, is the locus of the slider during one revolution of the crank.

It may be necessary to drive a secondary gear from the connecting rod, in which case the locus of the point P on the rod, from which the drive is taken (Fig. 1.13), will have to be determined. To plot this locus the crank and connecting rod should be drawn in, say, twelve indexing positions. Only four positions, O1, O3, O7 and O8,

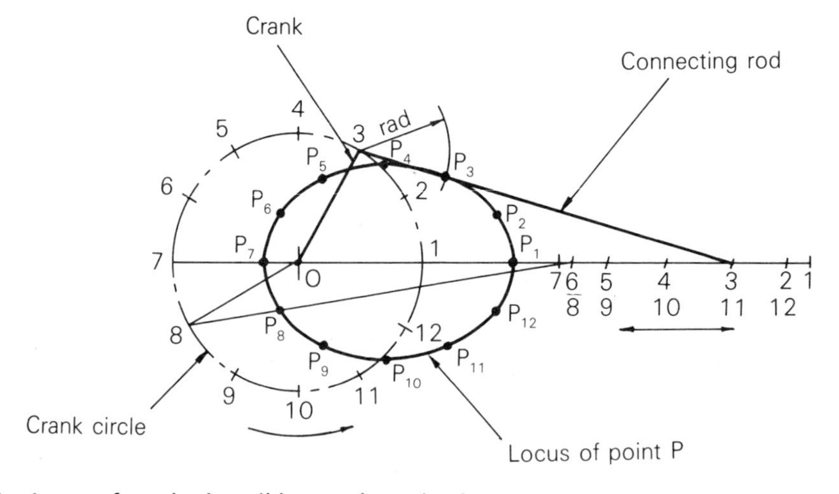

Fig. 1.13 Locus of a point in a slider-crank mechanism.

are shown in the diagram for simplification. With compasses set at the distance of P from the end of the crank, arcs are struck from successive positions 1, 2, 3, etc., on the crank circle, along the various connecting rod positions to give points P_1, P_2, P_3, etc. A smooth curve drawn through these points will give the locus.

Four-bar Chain Mechanism

A four-bar chain mechanism is shown in Fig. 1.14. AB and CD are links or bars which move about fixed pivots A and C. The two links are joined by a connecting link BD, which is pin-jointed at its ends (a pin-joint allows the connection to swivel). The fourth link, or bar, in the chain is the framework (not shown) which holds the pivots A and C a fixed distance apart. If a link EF is firmly attached to connecting link BD, then as AB and CD move to positions AB_1 and CD_1 respectively, EF will move to E_1F_1, the locus of point F being as shown during this movement. The locus is an approximate straight line and is known as Robert's straight-line mechanism.

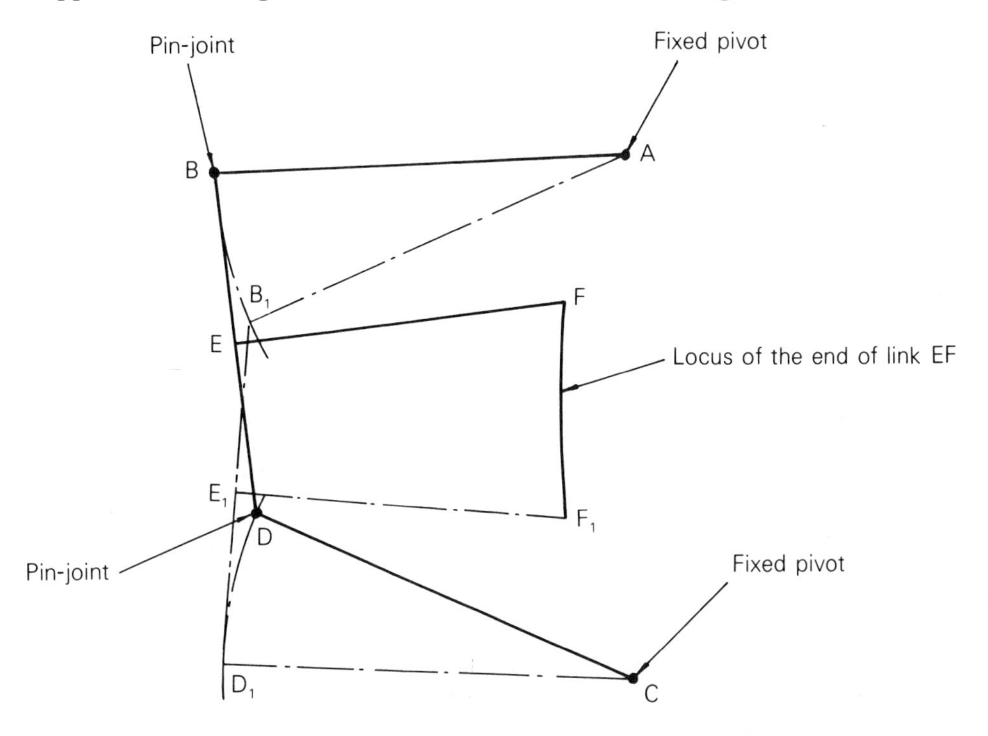

Fig. 1.14 Locus of a four-bar chain (Robert's straight-line mechanism)

Another four-bar chain mechanism is shown in Fig. 1.15. Links AB and CD oscillate from position AB_1 to AB_2 about centre A and from position CD_1 to CD_2 about centre C respectively. The links are joined by a connecting link BD, pin-jointed at its ends, which has a centre point P. The fourth link in the chain is the framework (not shown) which holds A and C at their fixed distance apart. To construct the locus of point P, draw in the various positions of AB, during oscillation, at, say, 15° intervals (this can easily be achieved using the 60°/30° and 45° set squares singly or combined). With compasses, set off the length BD from the successive positions of B on the arc B_1B_2 to cut the arc D_1D_2. Join the successive positions BD, etc., and mark on the centre point P. A smooth curve drawn through these points will give the locus. It can be seen that the locus is a straight line for part of the movement. This system, used by James Watt as a guide in his steam engines, is known as Watt's mechanism.

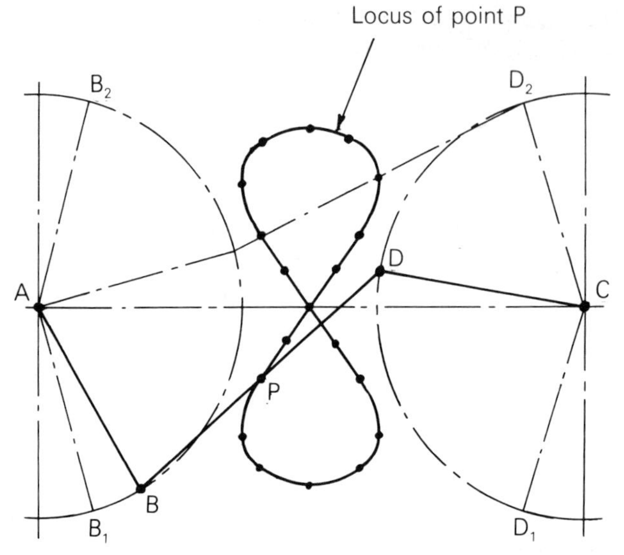

Fig. 1.15 Locus of a four-bar chain (Watt's mechanism).

SELF-TEST 1

Select the correct option(s). *Note*: There may be more than one correct answer.

1. To determine the locus of a point it is necessary to know its:

(a) starting position
(b) successive positions
(c) intermediate position
(d) finishing position.

2. On a bicycle, moving along a level road, a straight-line locus will be described by:

(a) the crank
(b) a point on the wheel rim
(c) the centre of a wheel hub
(d) a pedal.

3. As a roller rotates along a flat surface, a point on its periphery describes:

(a) a cycloid
(b) a straight line
(c) an ellipse
(d) a circle.

4. In Fig. 1.16, A and B are fixed points. Point T moves such that AT + BT is constant. The locus generated by point T is:

(a) a cycloid
(b) a straight line
(c) an ellipse
(d) a circle.

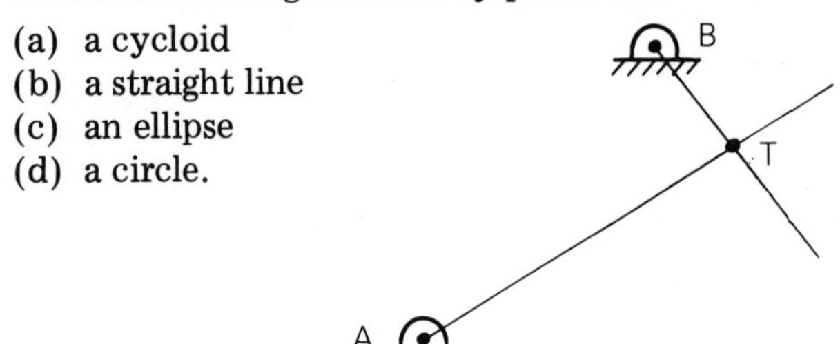

Fig. 1.16

5. In a simple chain-drive system the number of different types of locus generated by the chain connecting link is:

(a) one
(b) two
(c) three
(d) four.

6. A and B are two fixed points. Point P moves such that the distance AP is always equal to the distance BP. The resulting locus will be:

(a) a cycloid
(b) a straight line
(c) an ellipse
(d) a circle.

7. The types of loci generated by a crank pin and slider, in a slider–crank mechanism, are respectively:

(a) an ellipse and a straight line
(b) a straight line and a circle
(c) a straight line and an ellipse
(d) a circle and a straight line.

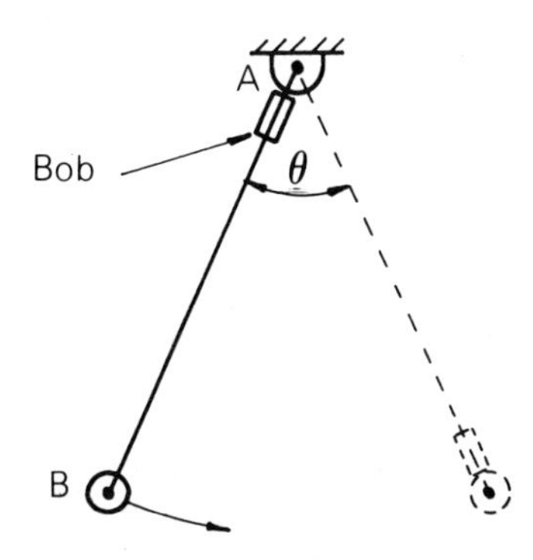

Fig. 1.17

8. Fig. 1.17 shows a pendulum AB which swings through angle θ. At the same time a bob, initially at A, moves uniformly down the pendulum and reaches B as the pendulum completes its swing. The resulting locus will be:

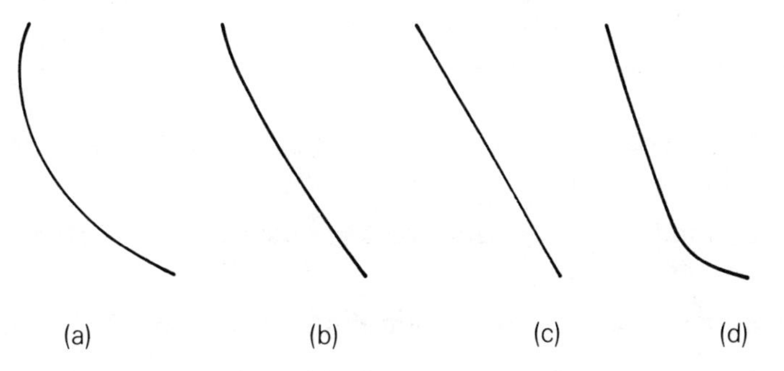

9. The principle of the quick return mechanism is used to:

(a) improve the efficiency of the working stroke
(b) give more flexibility
(c) improve the cycle time
(d) reduce the idle time.

10. Fig. 1.18 shows a wheel which, while rotating at a constant speed, moves from position X to position Y. Which of the following is correct?

I shows the locus of the centre of the wheel;
II shows the locus of a point on the rim:

(a) I only
(b) I and II
(c) II only
(d) neither I nor II.

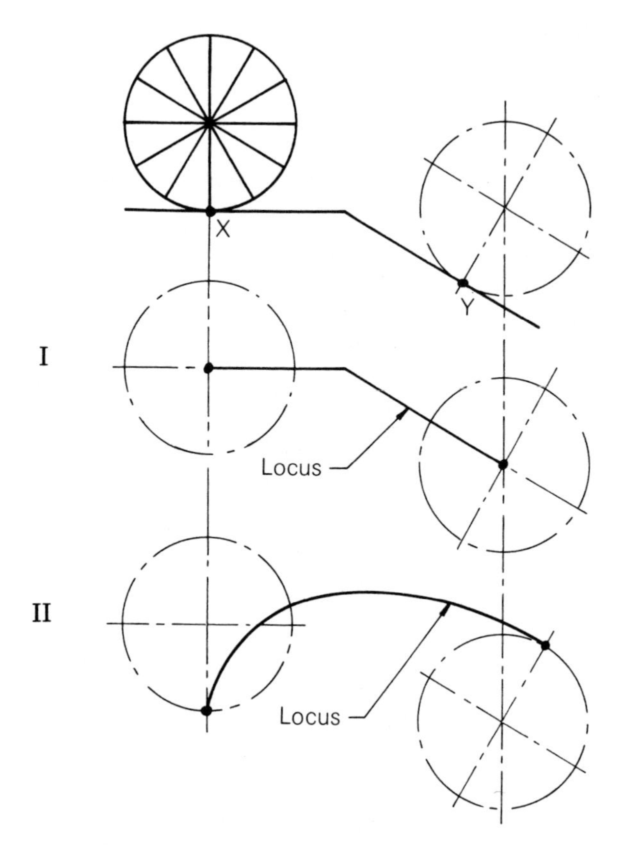

Fig. 1.18

EXERCISE 1

1. In your own words define the locus of a point. Support your definition with a suitable sketch.

2. Find the locus of a point P which is always 50 mm from a fixed straight line of infinite length.

3. Find the locus of a point P which is always 30 mm from a fixed point X.

4. A motor car which has tyres of 750 mm diameter, moves forward 1.77 m. Using a suitable scale, determine the locus of a point on the circumference of a tyre.

5. An overhead travelling crane lifts a gear blank at the rate of 0.5 m/s through a vertical height of 4 m. At the same time the crane travels a horizontal distance of 45 m. Using a suitable scale determine the locus of the gear blank.

6. X and Y are two fixed points 120 mm apart. Q is a moving point which is always 80 mm from X. R is a moving point which is always twice as far from Y as it is from X. Show the positions where Q and R coincide.

7. A wheel of 300 mm diameter moves at a controlled rate down two steps of tread 250 mm and fall 150 mm. Using a suitable scale determine the locus of the centre of the wheel.

8. A metal rod 150 mm long stands vertically against an angle plate which is mounted on a surface plate. The rod then slides so that its upper end describes a vertical line down the face of the angle plate and the lower end moves across the surface plate in a straight line at right angles to the angle plate. Draw the locus of the mid-point of the rod.

9. A quick return mechanism, similar to the one shown in Fig. 1.11, has a crank wheel of diameter 400 mm. A link L, 1 m long, pivots about point P which is 600 mm from the crank wheel centre. The other end of the link actuates a horizontal ram. Determine the movement of the ram during one complete revolution of the crank wheel. If the forward stroke is achieved in 3 s determine the time taken for the return stroke.

10. Using a suitable scale, determine the locus of the mid-point of the link L in Question 9 during one complete revolution of the crank wheel.

11. Fig. 1.19 shows a cylindrical bar of diameter 50 mm which has been cut off at an angle of 45°. Draw the resulting ellipse which would be seen when viewed in the direction of arrow A.

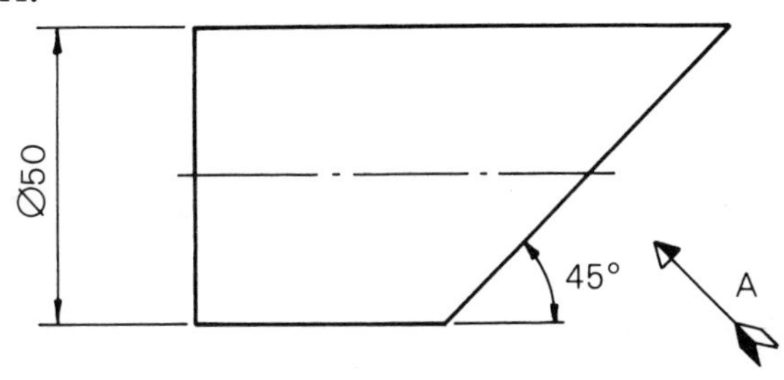

Fig. 1.19

12. Using different methods, construct an ellipse for each of the following sizes of major and minor axes:

(a) 75 mm × 40 mm
(b) 45 mm × 30 mm.

13. Fig. 1.20 shows a slider-crank mechanism. The crank OA rotates about fixed centre O. The connecting rod AP slides in a trunnion, which pivots about point X. If OA = 35 mm, AP = 130 mm and OX = 85 mm, construct the locus of point P.

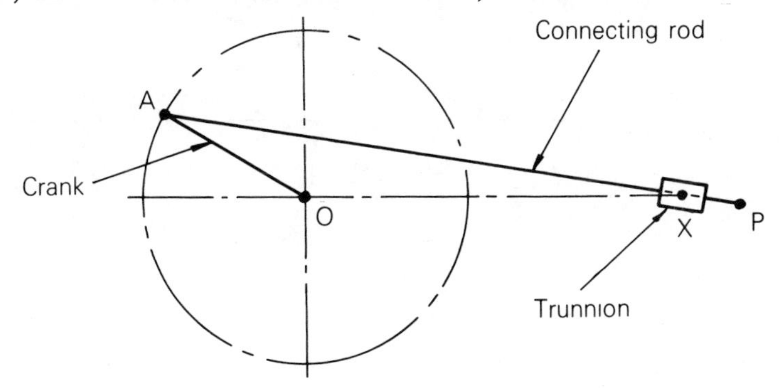

Fig. 1.20

14. Fig. 1.21 shows a crank mechanism with two cranks AB and CD, each 40 mm long, which oscillate in opposite directions about fixed centres A and C respectively. The two cranks are pin-jointed at B and D to a connecting link XY such that XB = YD = 60 mm and BD = 100 mm; centre distance AC = 120 mm. Plot the locus of point Y.

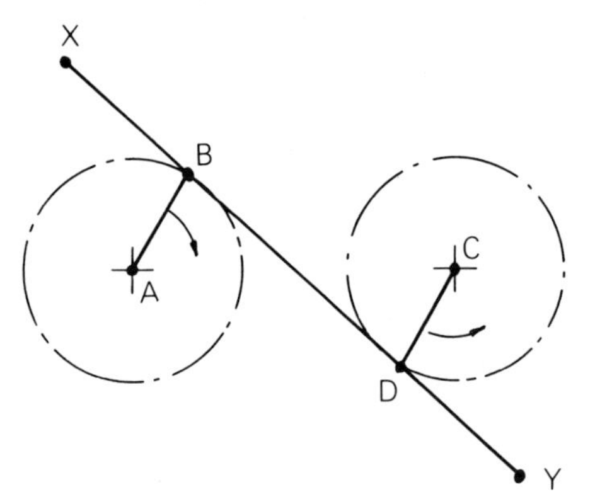

Fig. 1.21

15. A slider-crank mechanism is shown in Fig. 1.22 where the slider B acts in a line offset from the rotation of the crank OA, which revolves about fixed centre O. If OA = 40 mm and the connecting rod AB is 150 mm long, plot the locus of P, a point on the connecting rod, when AP = 60 mm.

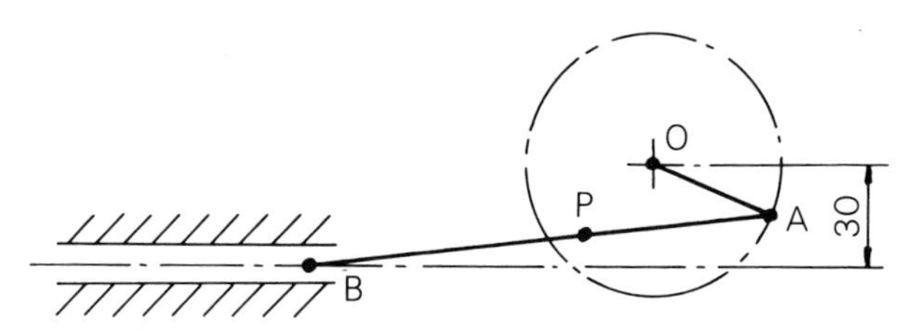

Fig. 1.22

16. The Scott-Russell mechanism is shown in Fig. 1.23. Crank OA rotates about fixed centre O and is pin-jointed to the link BD. The slider B is constrained to move in a straight line towards centre O. If AB = AD = OA = 50 mm, plot the locus of point D which will be seen to be a straight line for part of the motion.

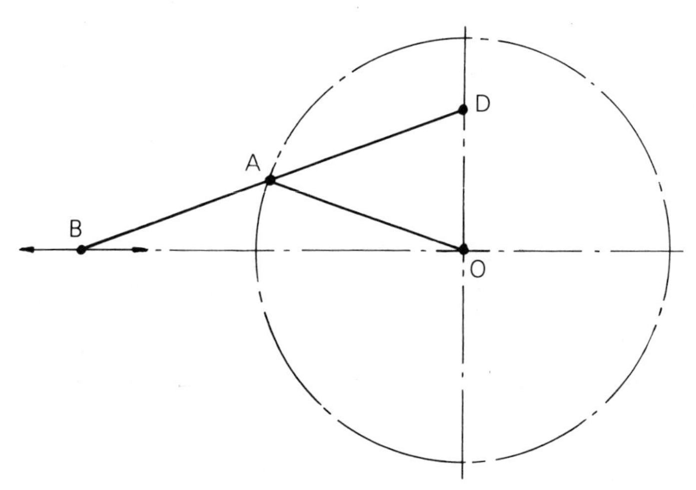

Fig. 1.23 Scott-Russell mechanism.

17. In Fig. 1.24 crank OA, which is 40 mm long, rotates about fixed centre O and causes crank CB to oscillate about fixed centre C through the connecting link AB. The mechanism is pin-jointed at A and B, and AB = 80 mm and BC = 60 mm. Plot the locus of point P, which is the mid-point of link AB.

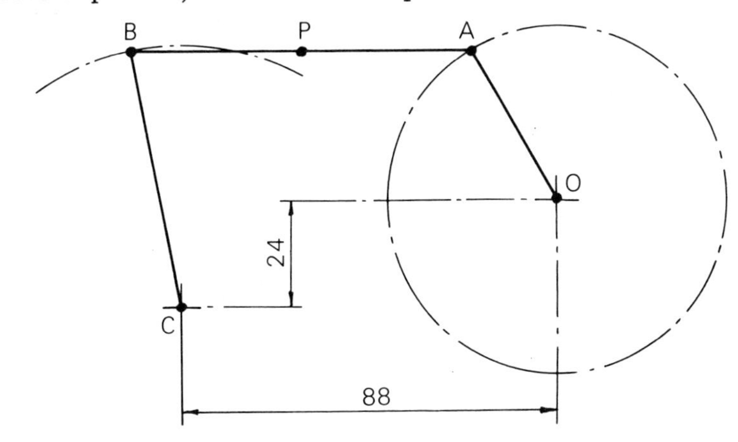

Fig. 1.24

18. A paint-dipping arrangement for tubular components is shown in Fig. 1.25. The tubes are suspended from the centre of the hanger bar between two sets of chain wheels. Draw the layout at a scale of 1:10 and determine the shape of a suitable guard which should be constructed to avoid danger from moving components and rotating chain wheels.

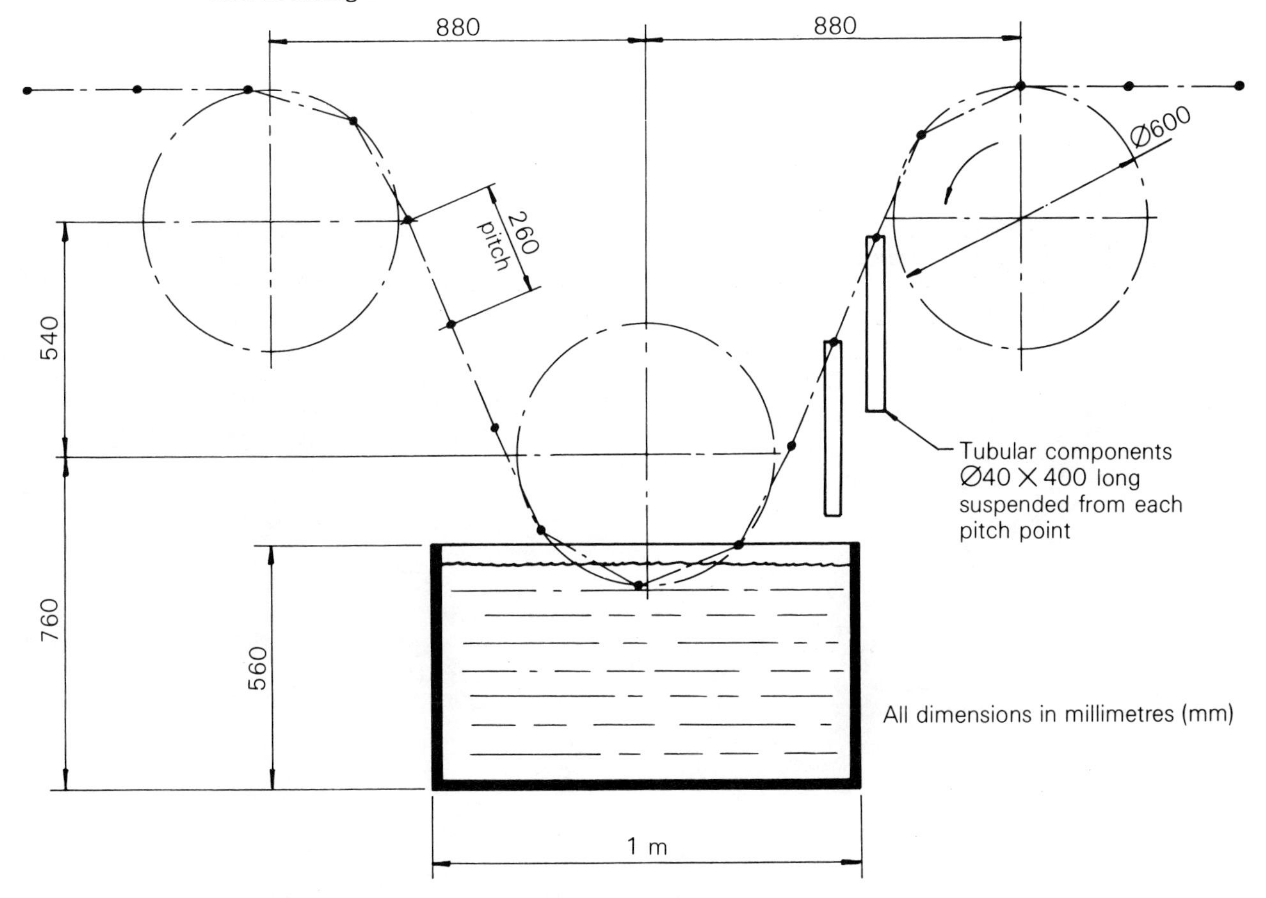

Fig. 1.25 Paint-dipping arrangement for tubes.

19. A schematic layout of a process conveyor for triangular architectural signs is shown in Fig. 1.26. The signs, which are in the form of an equilateral triangle of side 300 mm, are suspended from the centre of a hanger bar between two sets of chain wheels. Draw the layout at a scale of 1:10 and construct the shape of a guard necessary to shield the complete path of the signs while travelling from A to B.

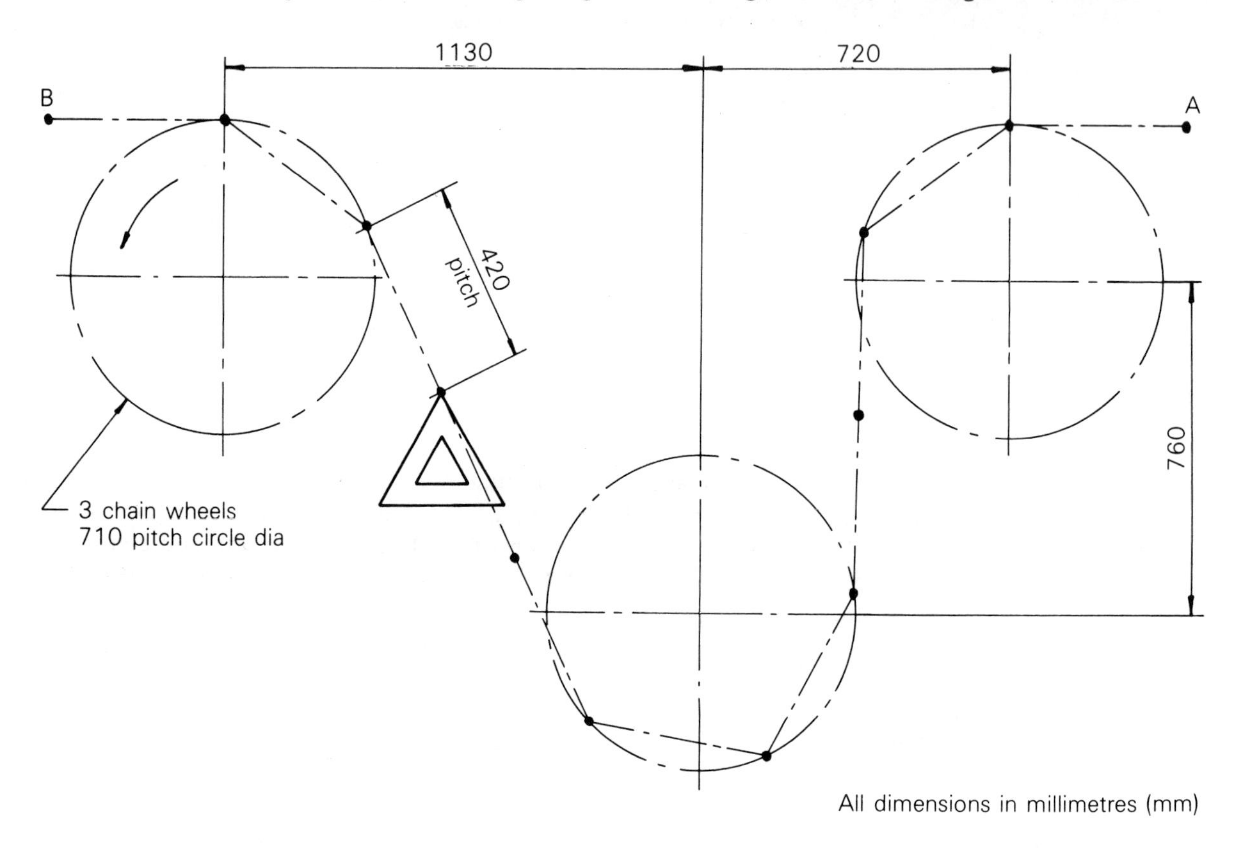

Fig. 1.26 Process conveyor for architectural signs.

SECTION B

CAMS

The Design and Function of Cams and Followers

After reading this chapter you should be able to:

- ★ appreciate the design and function of various cams and followers (G);
- ★ describe the function of a cam (S);
- ★ identify disc and cylindrical-type cams (S);
- ★ identify knife-edge, roller and flat followers (S);
- ★ relate cam profile to type of follower, motion of follower and cam movement (in-line, offset, radial arm) (S);
- ★ draw typical cam profiles for different followers and motions (S).

(G) = general TEC objective
(S) = specific TEC objective

THE FUNCTION OF A CAM

A cam is a component whose function is to provide some particular form of intermittent motion. It has a machined edge or groove which, during operation, maintains continuous contact with a follower. The edge or groove is so shaped as to cause the follower to move in the required manner as the cam is rotated.

Almost any reciprocating or intermittent motion can be imparted to the follower, and consequently cams are used for a great variety of engineering applications from the valve control in internal combustion engines to the tool control of automatic machine tools.

DISC AND CYLINDRICAL-TYPE CAMS

The direction of movement of the follower relative to the cam motion determines the basic type of cam system. A disc cam is one where the direction of movement of the follower is at right angles to the cam axis of rotation. A cylindrical cam is one where the follower movement is parallel to the cam axis.

Fig. 2.1 shows a disc cam. As the shaft rotates, the cam, which is rigidly fixed to it, also rotates and causes the follower to move perpendicular to the axis of its rotation.

Fig. 2.2 shows a cylindrical cam which is also mounted on to a shaft. As the cam rotates, the follower locates in the machined groove and moves in a direction parallel to the axis of the cam.

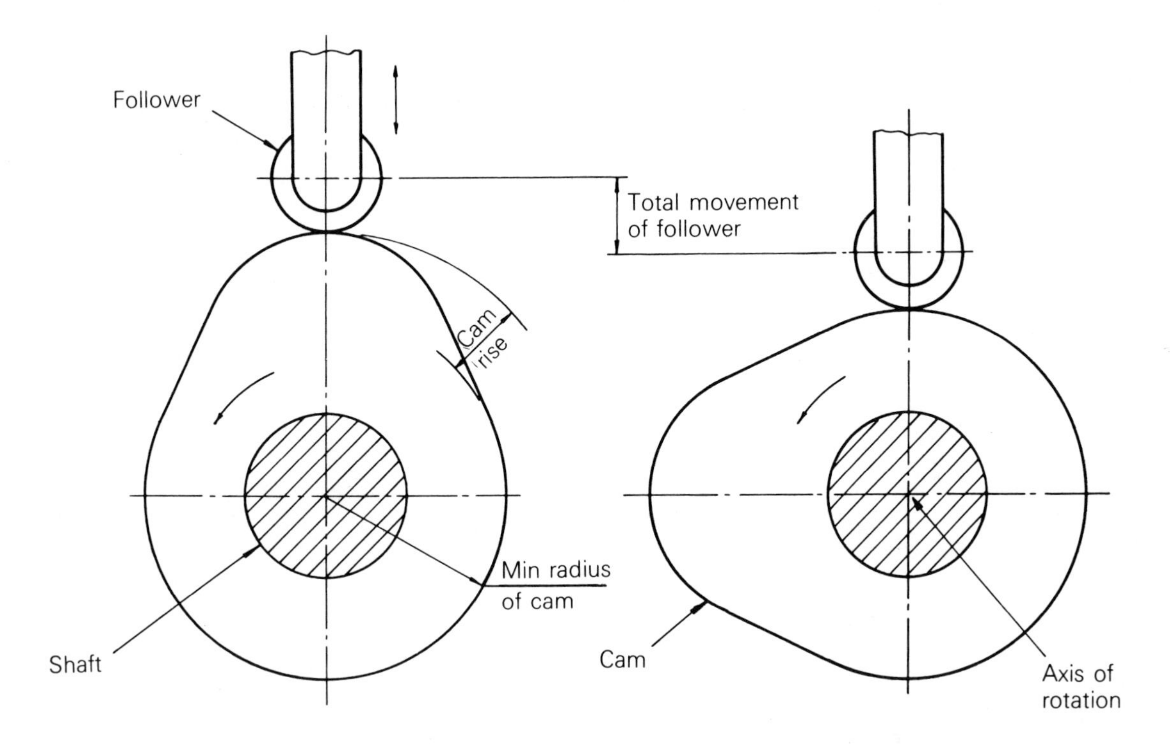

Fig. 2.1 Disc cam.

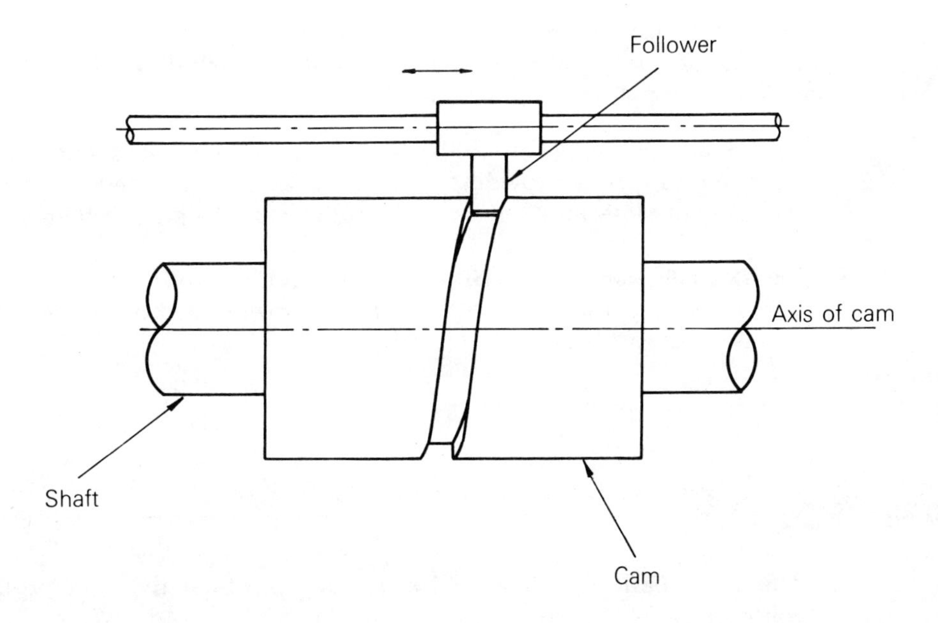

Fig. 2.2 Cylindrical cam.

TYPES OF FOLLOWER

The three main types of follower are identified relative to their shape and are shown in Fig. 2.3.

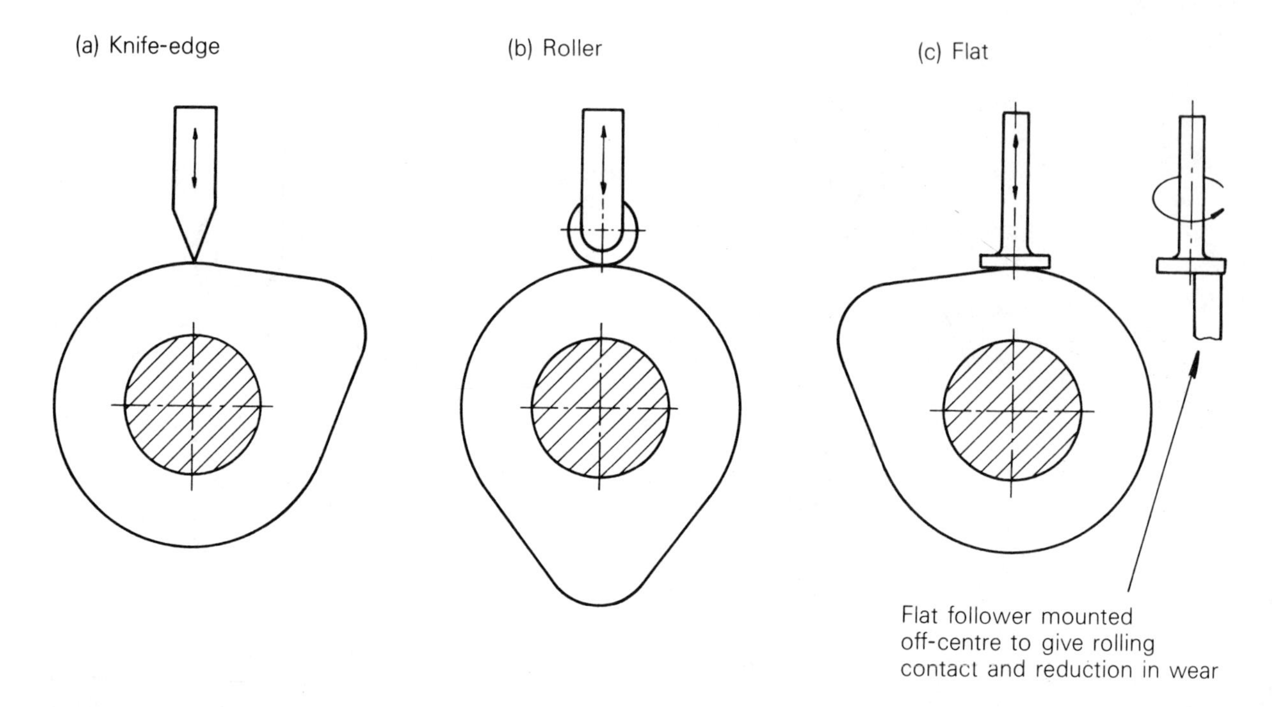

Fig. 2.3 Types of follower.

The knife-edge follower (Fig. 2.3a) has a knife edge which can follow intricate cam profiles, but by nature of its shape it will wear rapidly. It may be used in association with concave and convex cam profiles.

The roller follower (Fig. 2.3b) is most commonly used and gives a rolling action over the cam, thus providing a smooth action. Wear is reduced but the profile of the cam is limited because any concave radius must be greater than the roller radius.

The flat follower (Fig. 2.3c) is usually mounted off-centre in order to give rolling contact which results in less wear. Flat followers cannot be used in conjunction with concave cam profiles.

CAM PROFILES

There are basically three types of motion which the cam profile imparts to the follower:

(a) uniform velocity
(b) uniform acceleration
(c) simple harmonic motion.

For each of these motions it is the usual practice, when designing cam profiles, first to construct a displacement diagram which shows the rise and fall of the follower relative to the angular movement of the cam. Having drawn this diagram, the relative displacements may be transferred from it to construct the cam profile geometrically.

(a) Uniform Velocity

If a follower is to move with uniform velocity, it will travel at the same speed from the beginning to the end of its stroke. Because the movement starts from zero to maximum speed and stops in a similar abrupt way, there will be a sudden shock at the beginning and end of movement and the cam must be designed to remove this disturbance.

Fig. 2.4 shows a displacement diagram for a follower which is to rise with uniform velocity in 180° of a cam rotation. To construct a displacement diagram, a horizontal line is drawn (of any suitable length) to represent the angular movement of the cam and it is divided into, say, six equal parts and marked 0, 1, 2, etc. Vertical lines are constructed from each of these points, to a height equal to the rise of the cam, and they are also divided into the same number of equal parts and lettered A, B, C, etc. The intersections of the vertical and horizontal projections, P_1, P_2, P_3, etc., are joined to give a straight line and the diagram is completed.

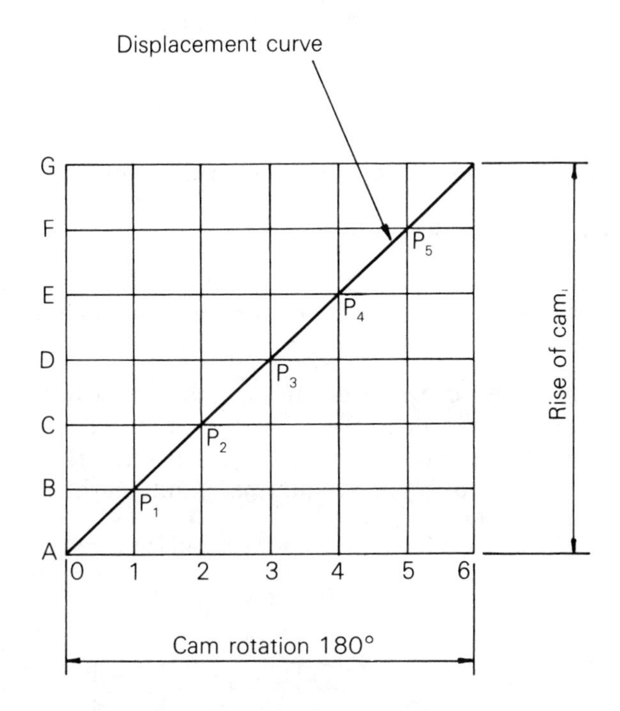

Fig. 2.4 Displacement diagram—uniform velocity.

(b) Uniform Acceleration

Where high speeds are involved it is better for a follower to start slowly and then accelerate at a uniform rate until maximum speed is reached. The follower is then decelerated (retarded) until the speed reaches zero and the follower has reached the end of its stroke.

Fig. 2.5 shows a displacement diagram for a follower which is to rise with uniform acceleration in 180° of a cam rotation. Again, horizontal and vertical lines are constructed in a similar way to the diagram for uniform velocity. Point O is joined to the junction of 3B, 3C and 3D and point X is joined to the junction of 3D, 3E and 3F. This construction produces the points P_1, P_2, P_3, etc., which are joined to complete the curve for uniform acceleration and retardation.

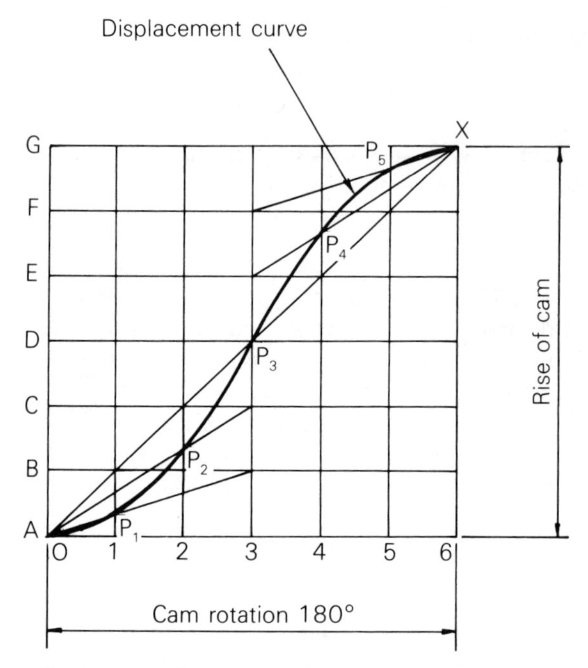

Fig. 2.5 Displacement diagram – uniform acceleration.

(c) Simple Harmonic Motion

Another system used for high-speed actions is simple harmonic motion where the follower is accelerated from zero and retarded back to zero to give a smoother action.

Fig. 2.6 shows a displacement diagram for a follower which is to rise with simple harmonic motion in 180° of cam rotation. To construct the diagram a horizontal line is drawn (of suitable length) to represent the radial movement of the cam and it is divided into, say, six equal parts and marked 0, 1, 2, etc. Vertical lines are con-

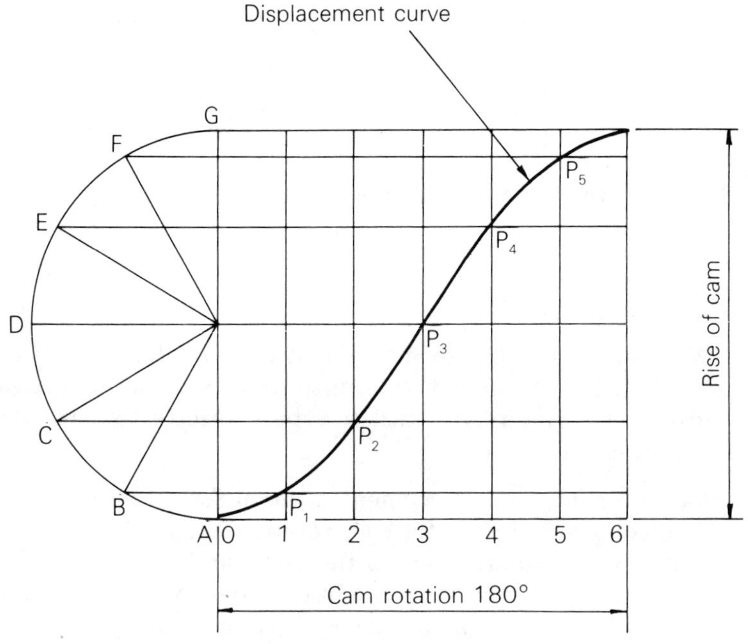

Fig. 2.6 Displacement diagram – simple harmonic motion.

structed from each of these points, to a height equal to the rise of the cam. A semi-circle is drawn from the mid-rise position and it is divided into the same number of angular parts as the horizontal line, in this case six, and marked A, B, C, etc. Horizontal lines are drawn from each of these positions to produce points P_1, P_2, P_3, etc., which are joined to produce the simple harmonic displacement curve. This is the shape of a sine curve.

Advantages may be gained by mounting followers in alternative positions relative to the cam rotation. Fig. 2.7a shows knife-edge and roller followers mounted in line with the axis of the cam. A smoother follower action may be obtained if the follower is offset from the cam axis. The offset is usually in the direction of motion. Fig. 2.7b shows offset knife-edge and roller followers. Because the rotation of the cam is anti-clockwise, the offset is to the left. Offsetting also has disadvantages, however, because side thrust on the follower is increased, which could result in bending the follower, and more side force is imparted on to the follower guides.

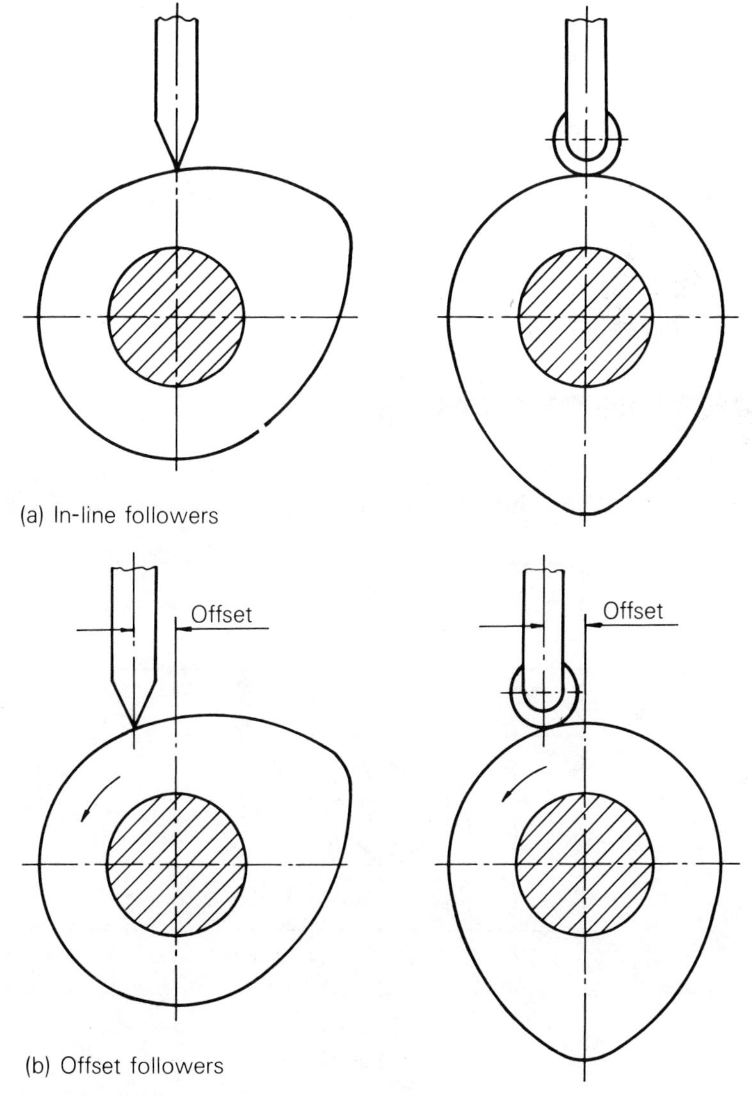

Fig. 2.7 In-line and offset followers.

It may be necessary to use the intermittent movement obtained from a cam at an angle and distance somewhat removed from its axis of rotation. This type of movement is frequently required in an internal combustion engine, an example of which is shown in Fig. 2.8. As the cam rotates it causes the follower to move and the rocker arm, which pivots about its fixed centre O, actuates the valve stem through the adjuster and valve cap. A compression spring, located under the cup, maintains contact between the cam and the follower.

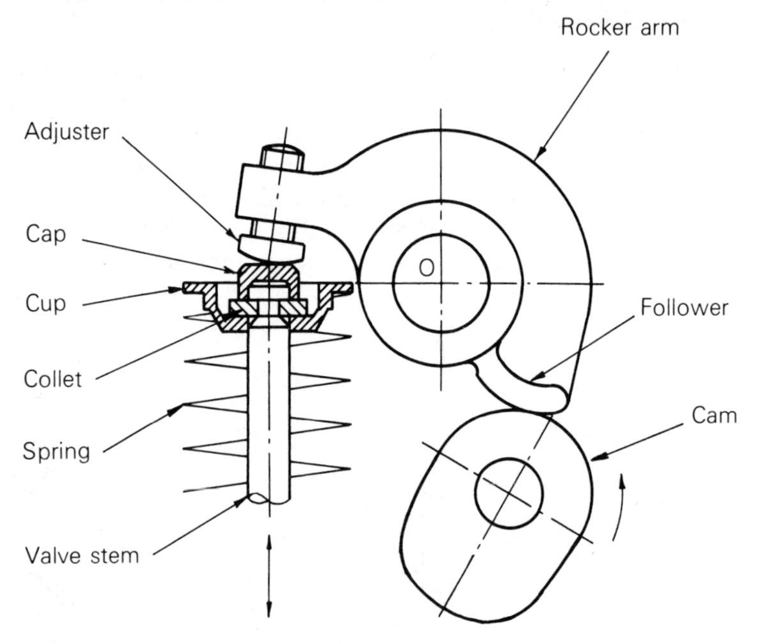

Fig. 2.8 Radial arm follower.

CONSTRUCTION OF CAM PROFILES

Example 1 A cam profile is required which will impart uniform velocity to an in-line knife-edge follower. The minimum radius of the cam is to be 45 mm. The follower is to rise

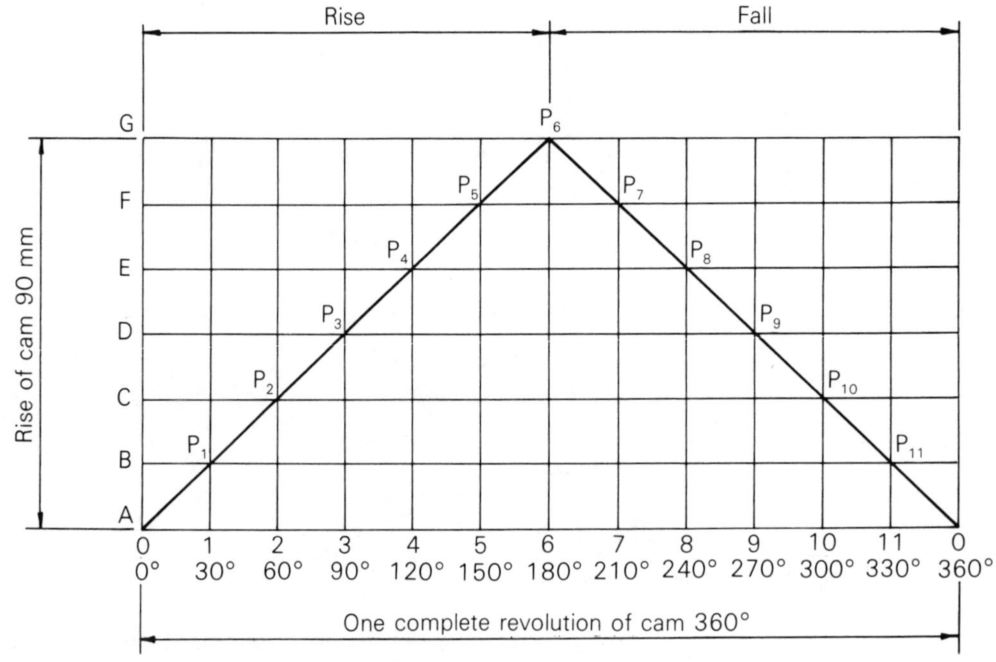

Fig. 2.9 Displacement diagram – uniform velocity.

90 mm in 180° of the cam rotation and fall 90 mm in the remaining 180° of rotation. Rotation of the cam is anti-clockwise.

Procedure

(a) Construct a uniform velocity displacement diagram (Fig. 2.9) by the method previously outlined. Because the rise and the fall of the follower take place during a similar amount of angular displacement of the cam, the displacement curve for the fall will be a mirror image of that for the rise.

(b) To construct the cam profile (Fig. 2.10) draw a circle, of radius 45 mm, to represent the minimum radius of the cam.

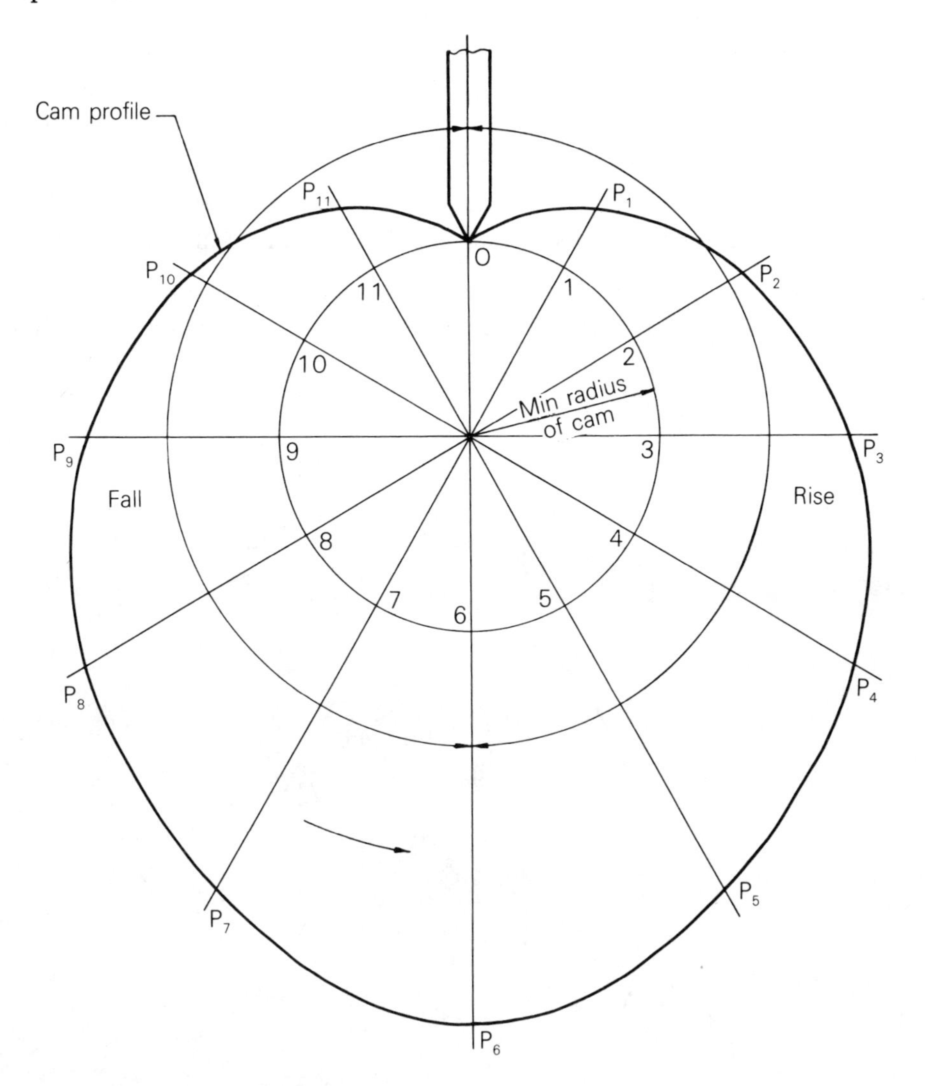

Fig. 2.10 Cam profile — uniform velocity.

(c) Starting at position O, divide the rise and fall into the same number of equal parts as the displacement diagram, i.e. six each (i.e. 30°), and mark them 1, 2, 3, etc. Draw radial lines through these points.

(d) On the displacement diagram, set dividers at the length of line $1P_1$ and transfer this to the cam profile drawing to locate the position of point P_1 from point 1 on the minimum radius circle.

(e) On the displacement diagram, set dividers at the length of line $2P_2$ and transfer this to the cam profile drawing to locate the position of point P_2 from point 2 on the minimum radius circle.

(f) Repeat the transfer of distances from the displacement diagram to obtain positions P_3, P_4, P_5, etc., on the cam profile drawing.

(g) Join up points O, P_1, P_2, P_3, etc., with a smooth curve to complete the cam profile.

Example 2 A cam profile is required which will impart uniform velocity to an in-line roller follower of diameter 25 mm. The minimum radius of the cam is to be 32.5 mm. All other details of the specification are the same as Example 1.

The profile necessary to transmit the same motion to a roller follower as in Example 1 is determined by the positions of the roller centre. Therefore it is necessary to plot the locus of the roller centre:

Minimum radius of cam (Example 1) = 45 mm

Minimum radius of cam + Radius of roller follower (Example 2) = 45 mm

Procedure (a) Construct a cam profile in a similar way to that produced in Example 1. This will be the locus of the roller centre and is shown chain-dotted (Fig. 2.11).

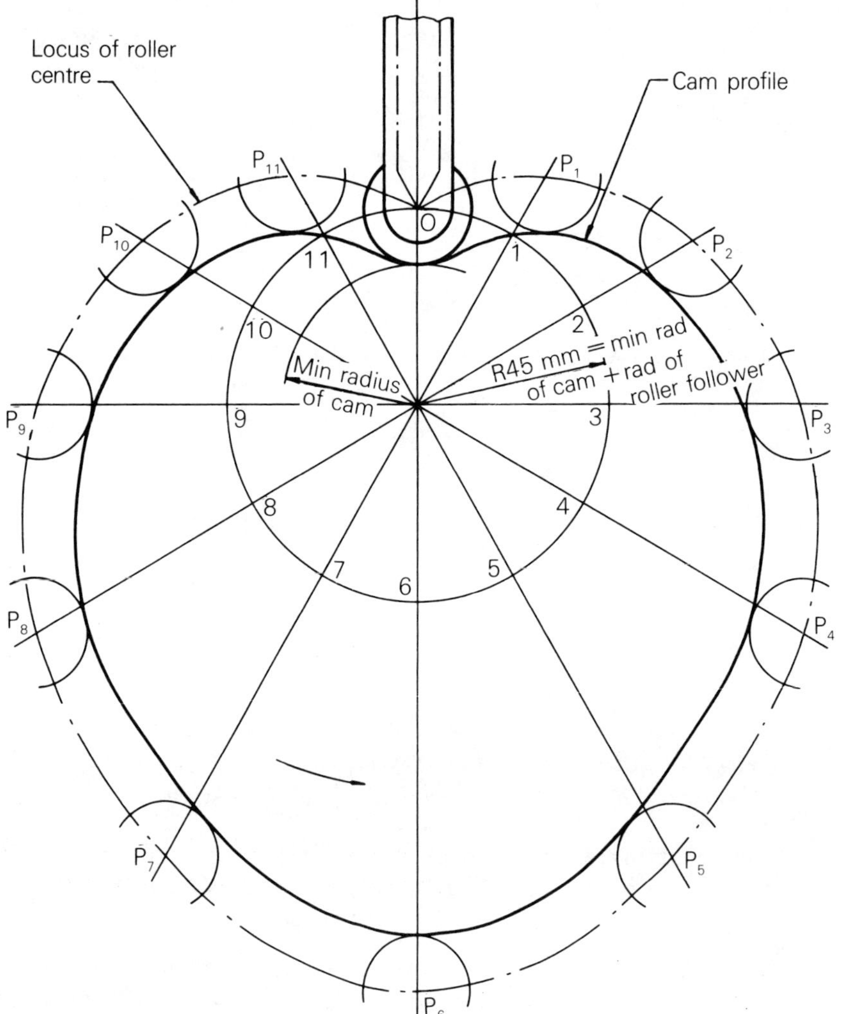

Fig. 2.11 Cam profile – uniform velocity.

(b) Set compasses at the roller radius, 12.5 mm, and draw a series of arcs, with centres O, P_1, P_2, P_3, etc., to represent the roller profile at these positions.

(c) Draw a smooth curve to touch the successive arcs to complete the cam profile.

Example 3 A cam profile is required which will impart uniform acceleration to an in-line knife-edge follower. The minimum radius of the cam is to be 45 mm. The follower is to rise 90 mm in 180° of the cam rotation and fall 90 mm in the remaining 180° of rotation. Rotation of the cam is anti-clockwise.

Procedure (a) Construct a uniform acceleration displacement diagram (Fig. 2.12) by the method previously outlined. Because the rise and the fall of the follower take place during a similar amount of angular displacement of the cam, the displacement curve for the fall will be a mirror image of that for the rise.

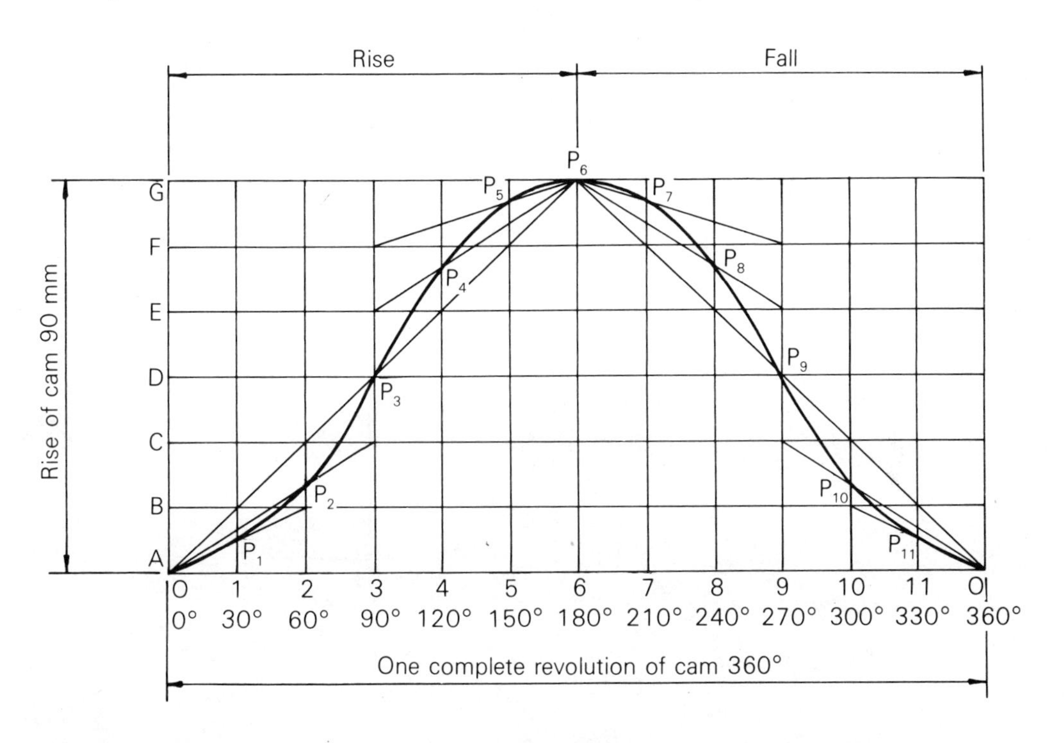

Fig. 2.12 Displacement diagram — uniform acceleration.

(b) To construct the cam profile (Fig. 2.13), draw a circle of radius 45 mm to represent the minimum radius of the cam.

(c) Starting at position O, divide the rise and fall into the same number of equal parts as the displacement diagram, i.e. six each (i.e. 30°), and mark them 1, 2, 3, etc. Draw radial lines through these points.

(d) On the displacement diagram, set dividers at the length of line $1P_1$ and transfer this to the cam profile drawing to locate the position of point P_1 from point 1 on the minimum radius circle.

(e) On the displacement diagram, set dividers at the length of line $2P_2$ and transfer this to the cam profile drawing to locate the position of point P_2 from point 2 on the minimum radius circle.

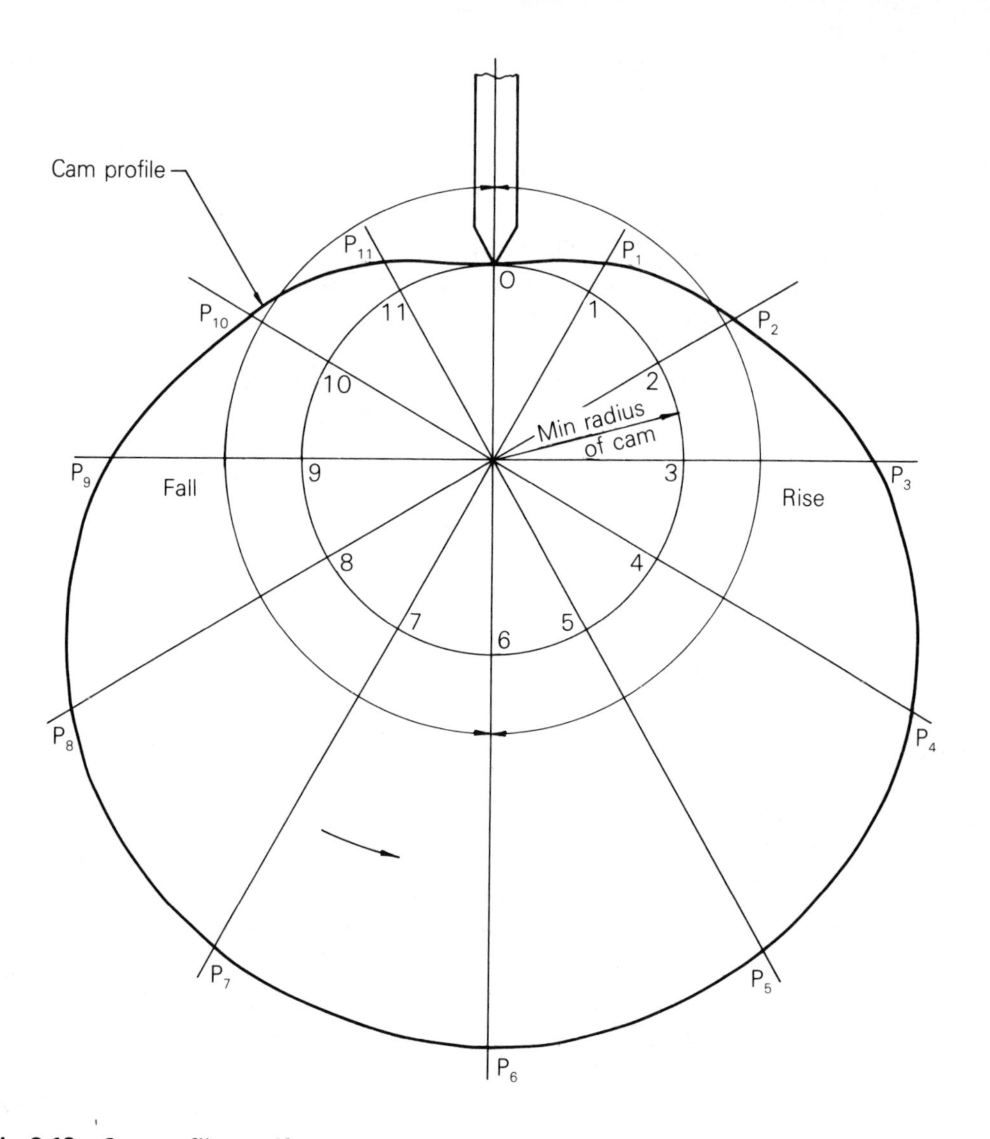

Fig. 2.13 Cam profile – uniform acceleration.

(f) Repeat the transfer of distances from the displacement diagram to obtain positions P_3, P_4, P_5, etc., on the cam profile drawing.

(g) Join up points O, P_1, P_2, P_3, etc., with a smooth curve to complete the cam profile.

Example 4 A cam profile is required which will impart uniform acceleration to an in-line roller follower of diameter 25 mm. The minimum radius of the cam is to be 32.5 mm. All other details of the specification are the same as Example 3.

The profile necessary to transmit the same motion to a roller follower as in Example 3 is determined by the positions of the roller centre. Therefore it is necessary to plot the locus of the roller centre:

$$\text{Minimum radius of cam (Example 3)} = 45\,\text{mm}$$

$$\text{Minimum radius of cam} + \text{Radius of roller follower (Example 4)} = 45\,\text{mm}$$

Procedure (a) Construct a cam profile in a similar way to that produced in Example 3. This will be the locus of the roller centre and is shown chain-dotted (Fig. 2.14).

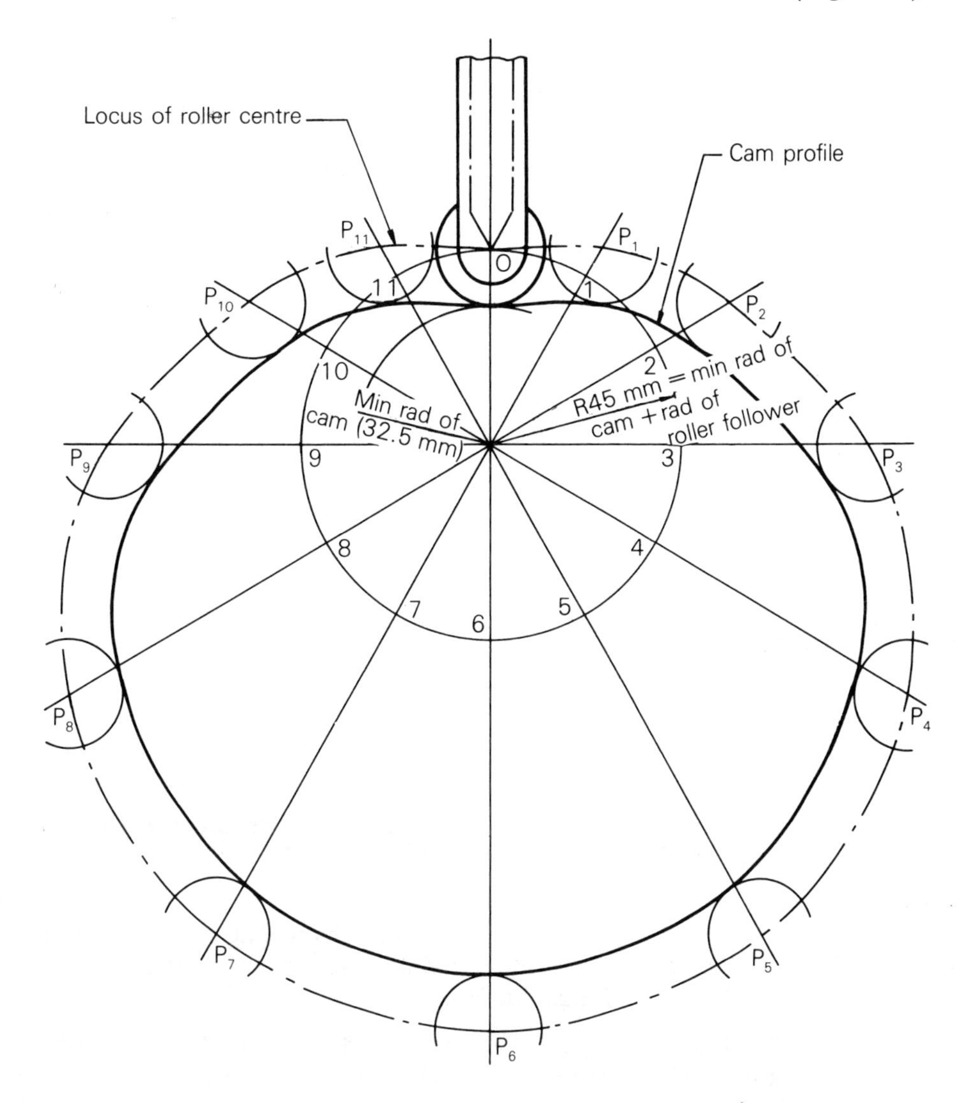

Fig. 2.14 Cam profile — uniform acceleration.

(b) Set compasses at the roller radius, 12.5 mm, and draw a series of arcs, with centres O, P_1, P_2, P_3, etc., to represent the roller profile at these positions.

(c) Draw a smooth curve to touch the successive arcs to complete the cam profile.

Example 5 A cam profile is required which will impart simple harmonic motion to an in-line knife-edge follower. The minimum radius of the cam is to be 45 mm. The follower is to rise 90 mm in 180° of the cam rotation and fall 90 mm in the remaining 180° of rotation. Rotation of the cam is anti-clockwise.

Procedure (a) Construct a simple harmonic motion displacement diagram (Fig. 2.15) by the method previously outlined. Because the rise and the fall of the follower take place during a similar amount of angular displacement of the cam, the displacement curve for the fall will be a mirror image of that for the rise.

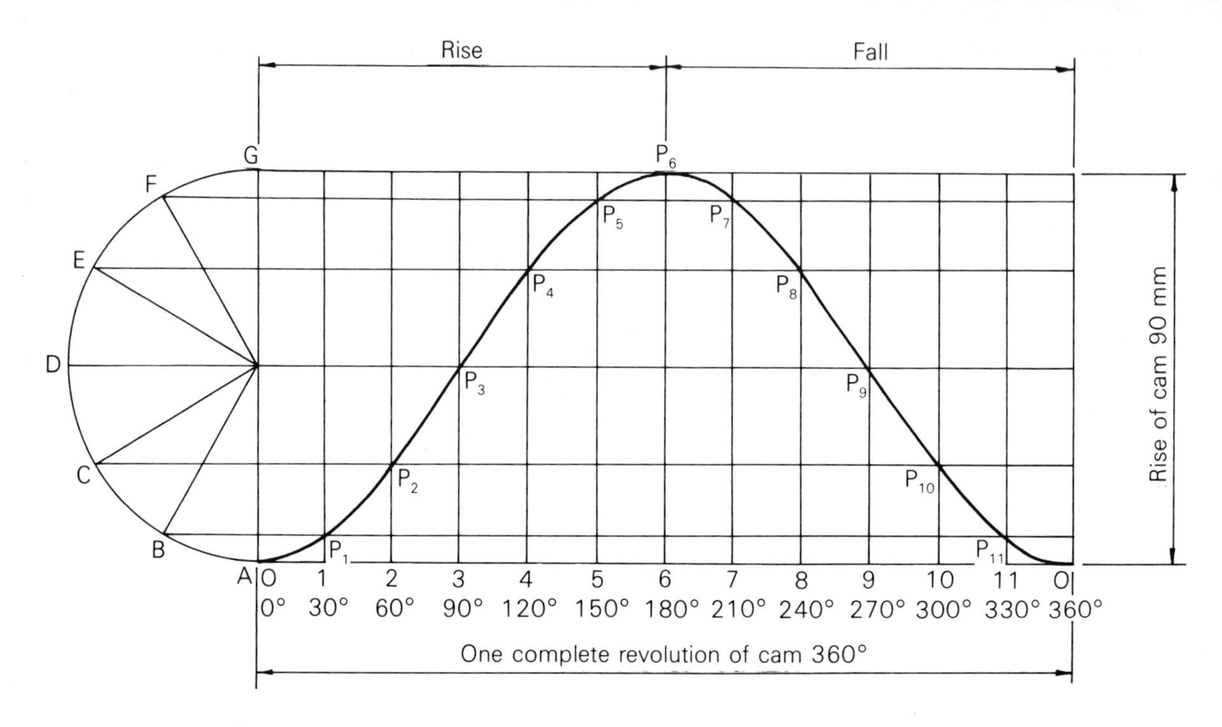

Fig. 2.15 Displacement diagram — simple harmonic motion.

(b) To construct the cam profile (Fig. 2.16), draw a circle of radius 45 mm to represent the minimum radius of the cam.

(c) Starting at position O divide the rise and fall into the same number of equal parts as the displacement diagram, i.e. six each (i.e. 30°), and mark them 1, 2, 3, etc. Draw radial lines through these points.

(d) On the displacement diagram, set dividers at the length of line $1P_1$ and transfer this to the cam profile drawing to locate the position of point P_1 from point 1 on the minimum radius circle.

(e) On the displacement diagram, set dividers at the length of line $2P_2$ and transfer this to the cam profile drawing to locate the position of point P_2 from point 2 on the minimum radius circle.

(f) Repeat the transfer of distances from the displacement diagram to obtain positions P_3, P_4, P_5, etc., on the cam profile drawing.

(g) Join up points O, P_1, P_2, P_3, etc., with a smooth curve to complete the cam profile.

Example 6 A cam profile is required which will impart simple harmonic motion to an in-line roller follower of diameter 25 mm. The minimum radius of the cam is to be 32.5 mm. All other details of the specification are the same as Example 5.

The profile necessary to transmit the same motion to a roller follower as in Example 5 is determined by the positions of the roller centre. Therefore it is necessary to plot the locus of the roller centre:

$$\text{Minimum radius of cam (Example 5)} = 45\,\text{mm}$$

$$\text{Minimum radius of cam} + \text{Radius of roller follower (Example 6)} = 45\,\text{mm}$$

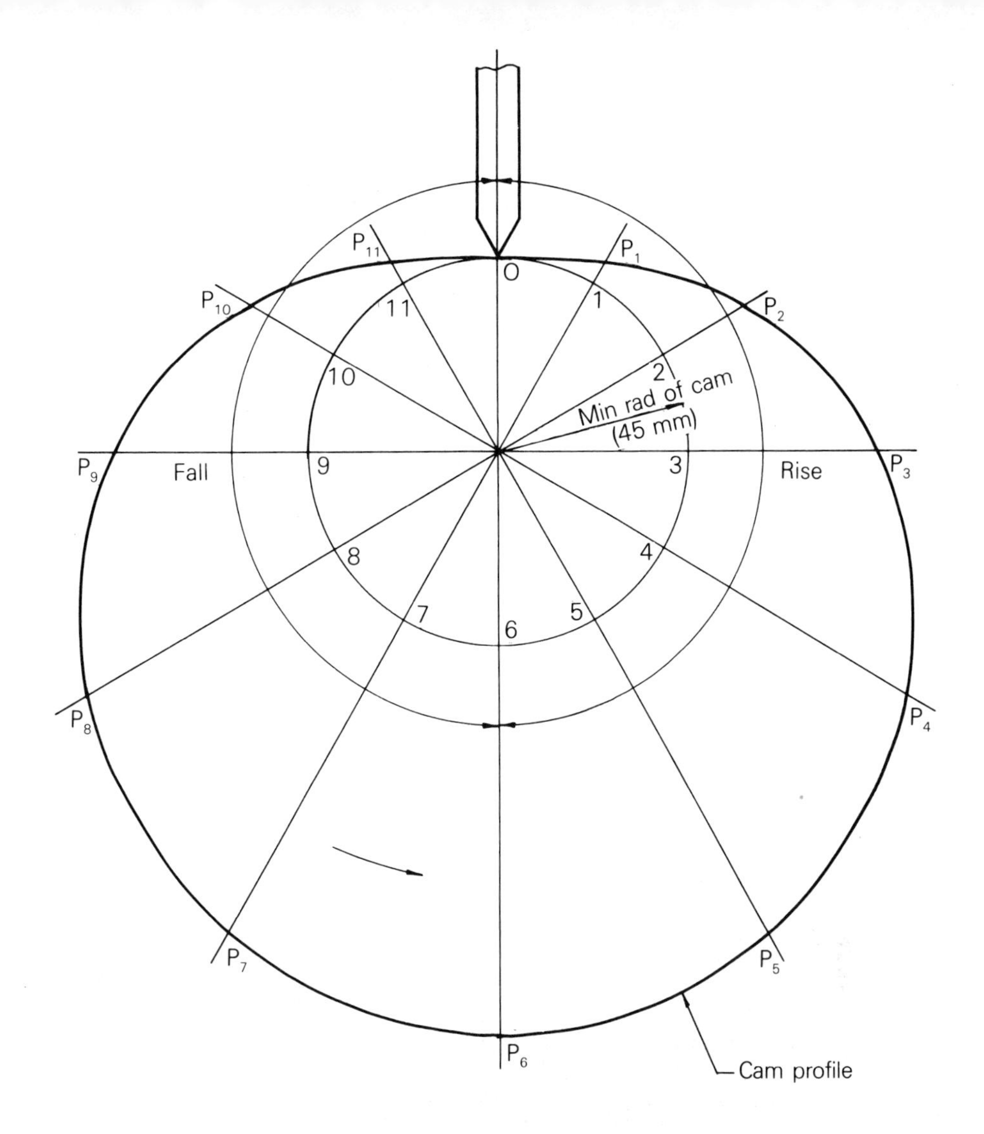

Fig. 2.16 Cam profile — simple harmonic motion.

Procedure

(a) Construct a cam profile in a similar way to that produced in Example 5. This will be the locus of the roller centre and is shown chain-dotted (Fig. 2.17).

(b) Set compasses at the roller radius, 12.5 mm, and draw a series of arcs, with centres O, P_1, P_2, P_3, etc., to represent the roller profile at these positions.

(c) Draw a smooth curve to touch the successive arcs to complete the cam profile.

Example 7 A cam profile is required which will impart uniform acceleration to an offset roller follower of diameter 25 mm. The minimum radius of the cam is to be 32.5 mm. The follower, which is offset 20 mm to the left, is to rise 90 mm in 180° of cam rotation and fall 90 mm in the remaining 180° of rotation. Rotation of the cam is anti-clockwise.

Procedure

(a) Construct a uniform acceleration displacement diagram (Fig. 2.12).

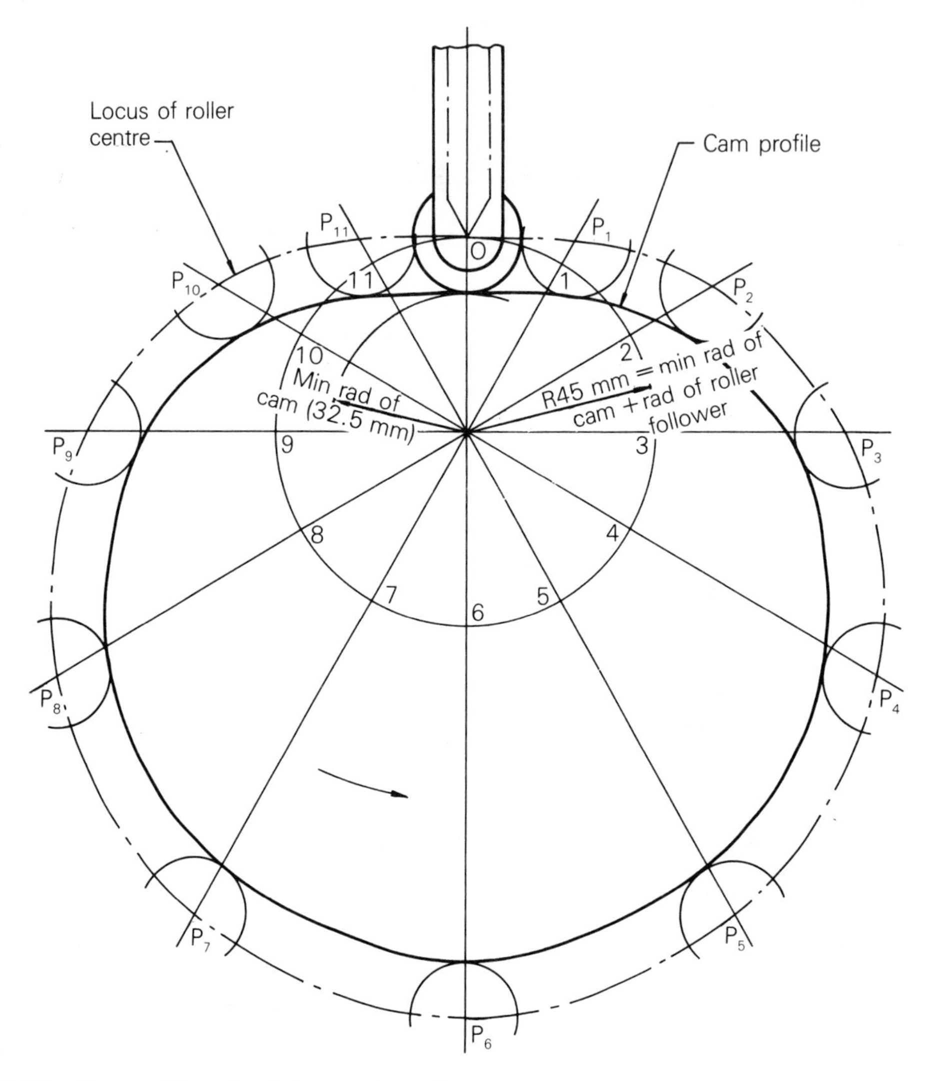

Fig. 2.17 Cam profile – simple harmonic motion.

(b) To construct the cam profile (Fig. 2.18), draw a circle of radius 45 mm to represent the minimum radius of the cam + the radius of the roller follower.

(c) Draw a circle of radius equal to the offset amount, i.e. 20 mm, concentric with the minimum cam radius circle.

(d) Divide the R20 circle into twelve equal parts (i.e. 30° each) and construct tangents to cut the minimum cam radius circle at positions 1, 2, 3, etc.

(e) On the displacement diagram, set dividers at the length $1P_1$ and transfer this to the cam profile drawing to locate the position of point P_1 from point 1 on the minimum radius circle.

(f) Repeat the transfer of distances from the displacement diagram to obtain positions P_2, P_3, P_4, etc., on the cam profile drawing.

(g) If the points O, P_1, P_2, P_3, etc., are joined with a smooth curve, it will complete the profile for a knife-edge follower (shown here chain-dotted). In this case it is the locus of the roller centre and can be used to obtain the arcs which represent the successive positions of the roller during the cam rotation.

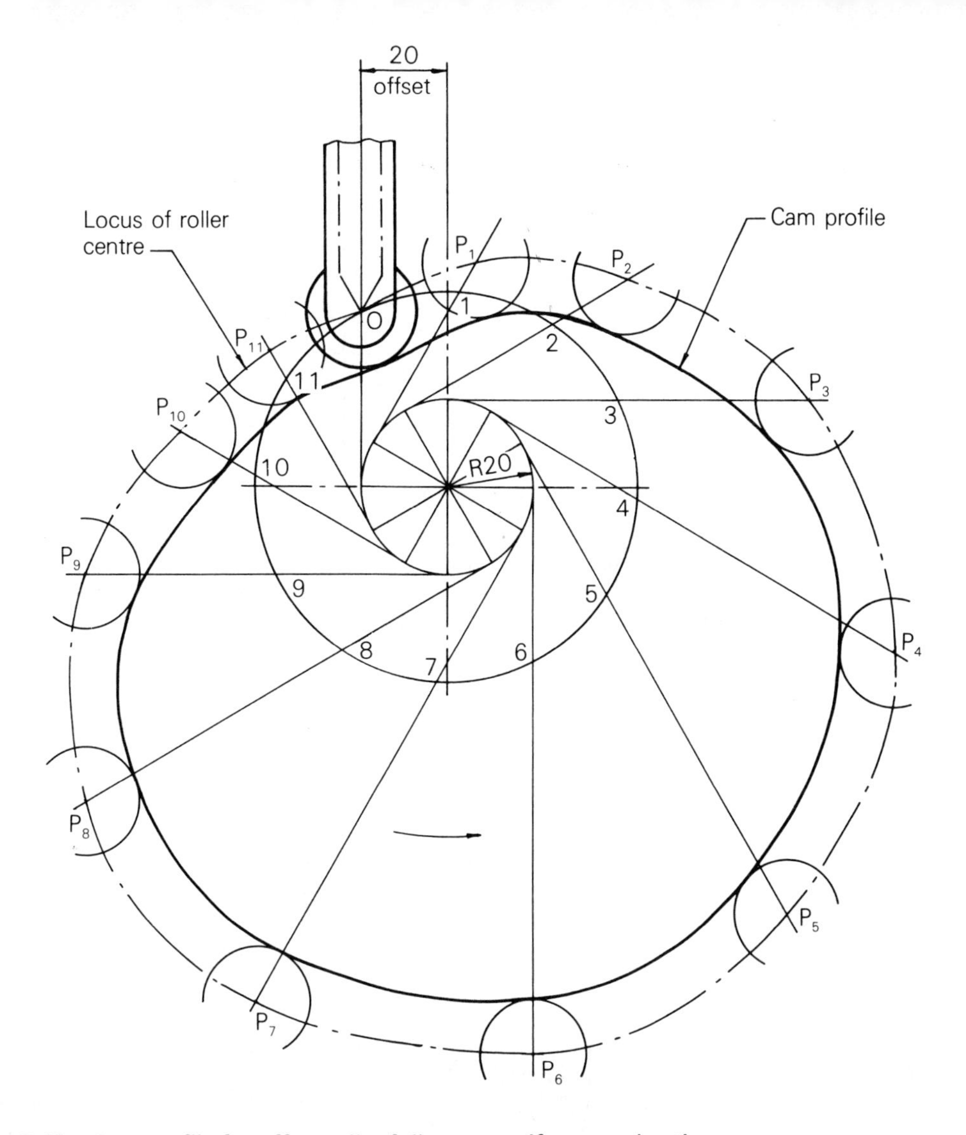

Fig. 2.18 Cam profile for offset roller follower — uniform acceleration.

(h) Construct a series of arcs with centres P_1, P_2, P_3, etc., as in previous examples.

(i) Draw a smooth curve to touch the arcs to complete the cam profile.

Example 8 A cam profile is required which will impart uniform velocity to an in-line knife-edge follower under the following conditions:

minimum radius of cam 32 mm
the follower to:
rise 50 mm in 90° of cam rotation
dwell for 90° of cam rotation
fall 50 mm in 180° of cam rotation
rotation of cam anti-clockwise.

Procedure (a) Construct a displacement diagram (Fig. 2.19) for the specification. A dwell action is one where the follower remains in a stationary position during part of the cam rotation. On a displacement diagram this action will be represented by a straight line, parallel to the base line.

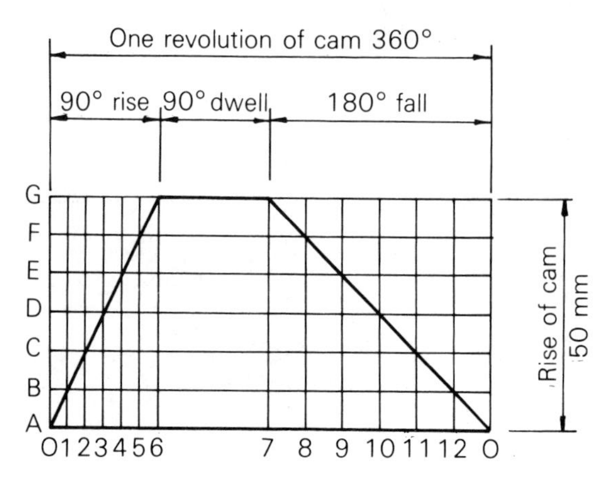

Fig. 2.19 Displacement diagram — uniform velocity.

(b) Construct the cam profile by the method outlined in previous examples (Fig. 2.20). For the period of the dwell of the follower the cam profile will be a radius struck from the cam centre of rotation X.

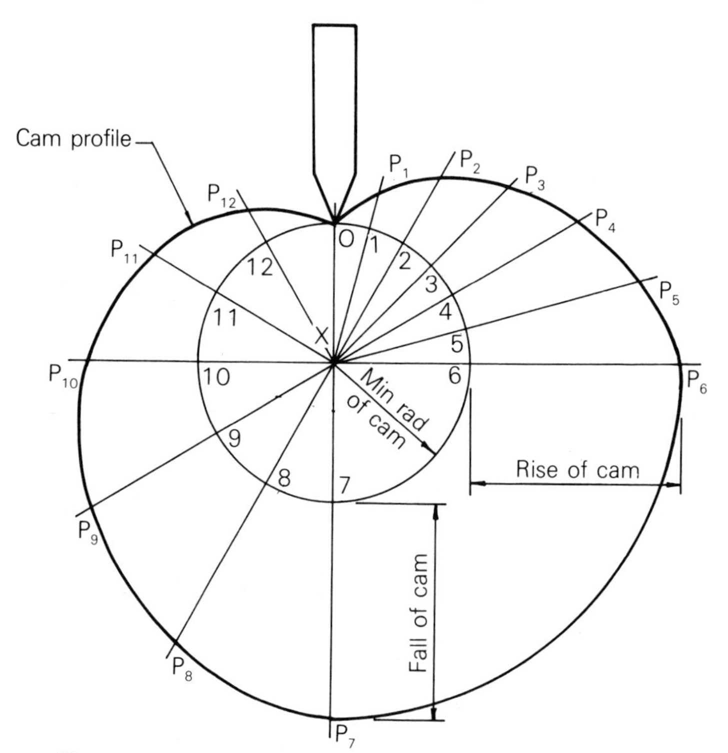

Fig. 2.20 Cam profile.

Example 9 A cam profile is required which will impart simple harmonic motion to an in-line roller follower under the following conditions:

minimum radius of cam 30 mm
the follower (20 mm diameter) to:
- rise 30 mm in 90° of cam rotation
- dwell for 30° of cam rotation
- fall 30 mm in 180° of cam rotation
- dwell for 60° of cam rotation

rotation of cam anti-clockwise.

Procedure (a) Construct a displacement diagram (Fig. 2.21) for the specification.

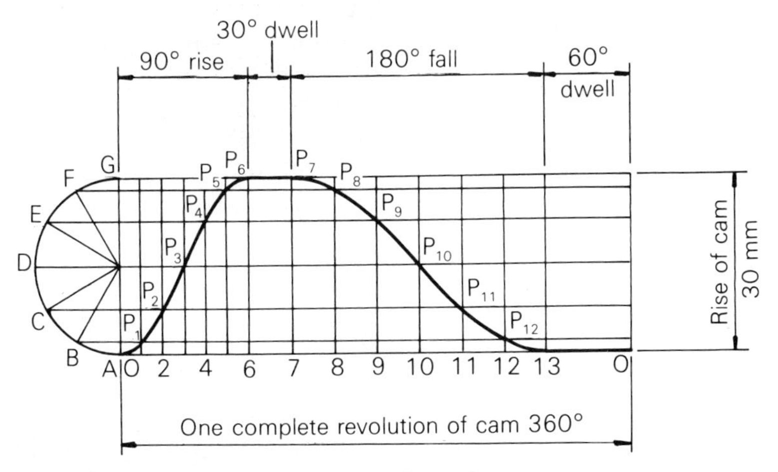

Fig. 2.21 Displacement diagram – simple harmonic motion.

(b) Construct the cam profile by the method outlined in previous examples (Fig. 2.22).

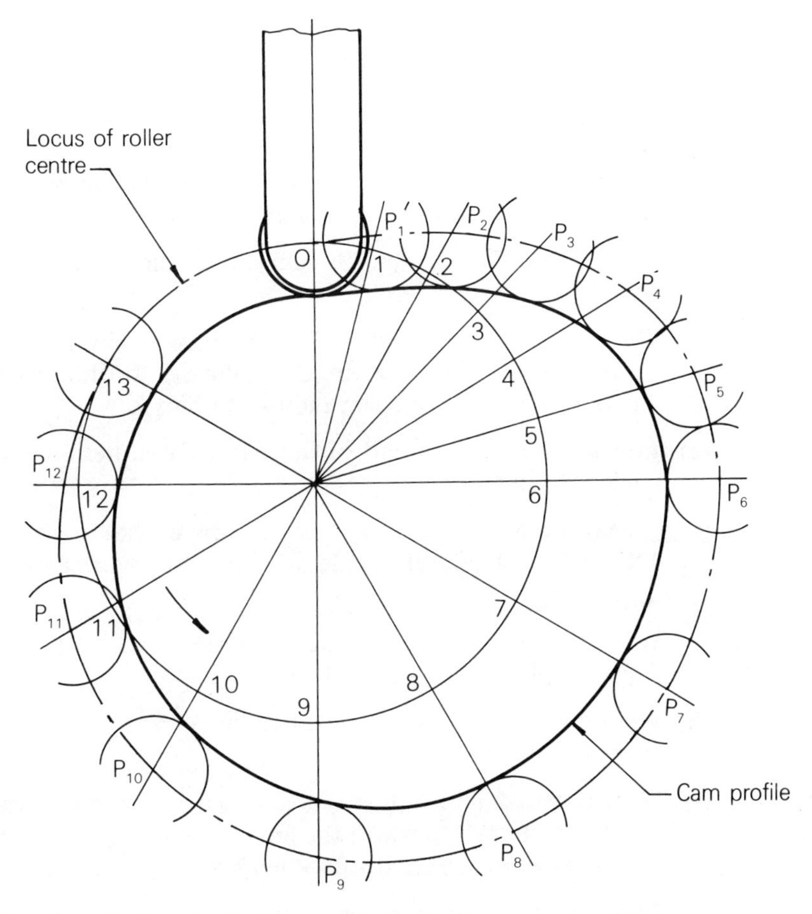

Fig. 2.22 Cam profile.

Example 10 A cam profile is required which will impart simple harmonic motion to a radial arm follower, similar to the example shown in Fig. 2.8, under the following conditions:

minimum radius of cam 40 mm
the follower to:
dwell for 90° of cam rotation
rise 20 mm in 75° of cam rotation
dwell for 30° of cam rotation
fall 20 mm in 75° of cam rotation
dwell for 90° of cam rotation
rotation of cam anti-clockwise.

Procedure

(a) Construct a displacement diagram (Fig. 2.23) for the specification.

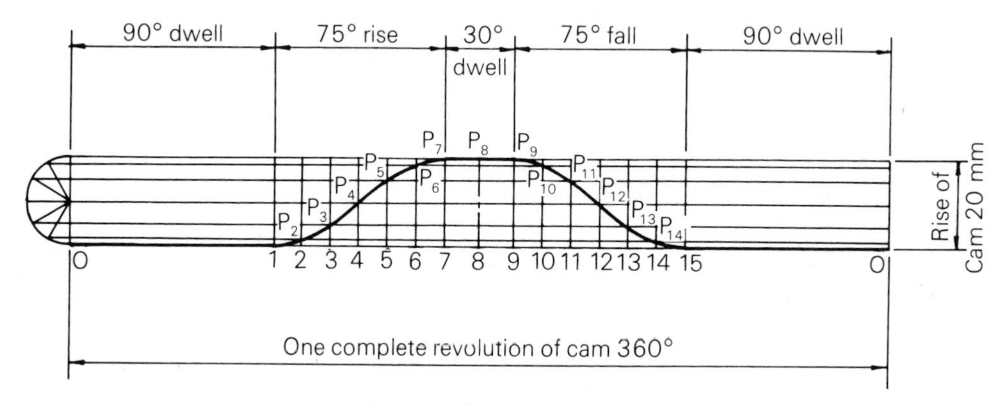

Fig. 2.23 Displacement diagram — simple harmonic motion.

(b) To construct the cam profile (Fig. 2.24), draw a circle of radius 40 mm to represent the minimum radius of the cam.

(c) With centre X and radius 99 mm draw a circle to pass through the follower pivot O.

(d) With centre X and radius 75 mm draw a circle to pass through the follower profile centre Y. Draw in the follower profile with the arc radius, 35 mm, meeting the minimum cam radius at point Z.

(e) Locate point 1 (the end of the initial dwell period) on the pivot circle such that angle OX1 is 90°.

(f) Locate point 7 (the end of the follower rise) on the pivot circle such that angle 1X7 is 75°. Divide this angle into six equal parts and label them, 2, 3, 4, etc.

(g) With centre 1 and radius OY mark off point P_1 on the follower profile circle. With centre P_1 and radius 35 mm draw in an arc to represent the position of the follower profile after 90° of cam rotation.

(h) With centre X draw an arc of radius (75 mm + distance $2P_2$ from the displacement diagram).

(i) With centre 2 and radius OY draw an arc to cross the arc drawn in (h) at point P_2. With centre P_2 draw in the arc (R35) to represent the position of the follower profile after one-sixth displacement of the cam rise.

(j) With centre X draw an arc of radius (75 mm + distance $3P_3$ from the displacement diagram).

(k) With centre 3 and radius OY draw an arc to cross the arc drawn in (j) at point P_3. With centre P_3 draw in the arc (R35) to represent the position of the follower profile after two-sixths (one-third) displacement of the cam rise.

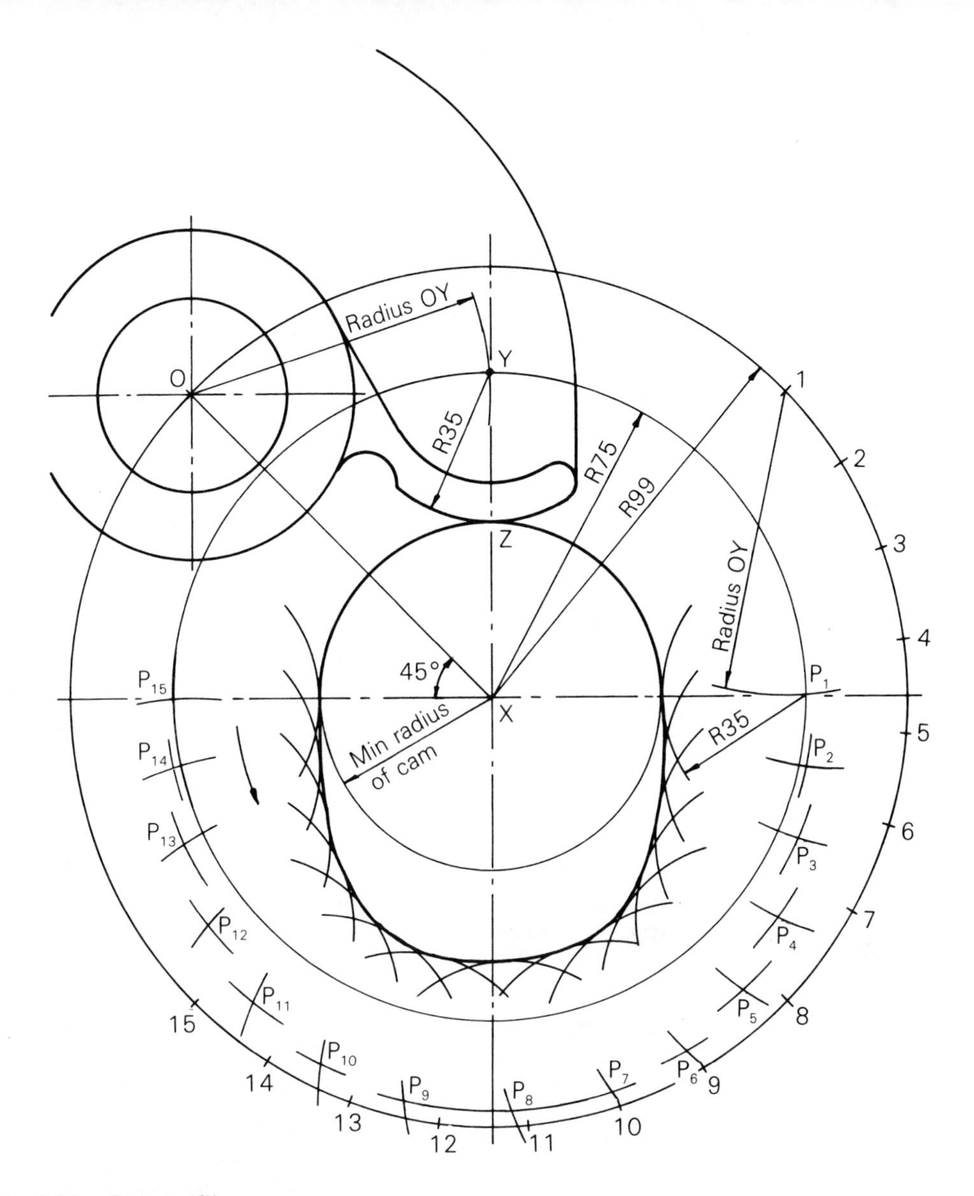

Fig. 2.24 Cam profile.

(l) Draw in the positions of profile arcs with centres P_3 to P_7 inclusive by the method outlined in (g) to (k).

(m) Locate point 9 (the end of the second dwell period) on the pivot circle such that angle 7X9 is 30° and bisect this angle to give point 8.

(n) With centre X draw an arc of radius (75 mm + distance $8P_8$ from the displacement diagram).

(o) With centres 8 and 9 and radius OY draw arcs to cross the arc drawn in (n) at points P_8 and P_9 respectively. With centres P_8 and P_9 draw in the arcs (R35) to represent the position of the follower profile during the 30° dwell period.

(p) Follow a similar method to draw in the profile arcs for the remainder of the cam rotation.

(q) Draw a smooth curve to touch the successive arcs to complete the cam profile.

Example 11 A cylindrical cam profile is required which will impart simple harmonic motion to a roller follower under the following conditions:

cam diameter 60 mm
cam rise 48 mm
roller follower diameter 8 mm
depth of groove 4 mm
the follower to:
rise 48 mm in 90° of cam rotation
dwell for 180° of cam rotation
fall 48 mm in 90° of cam rotation.

Procedure (a) Construct a displacement diagram (Fig. 2.25) for the specification.

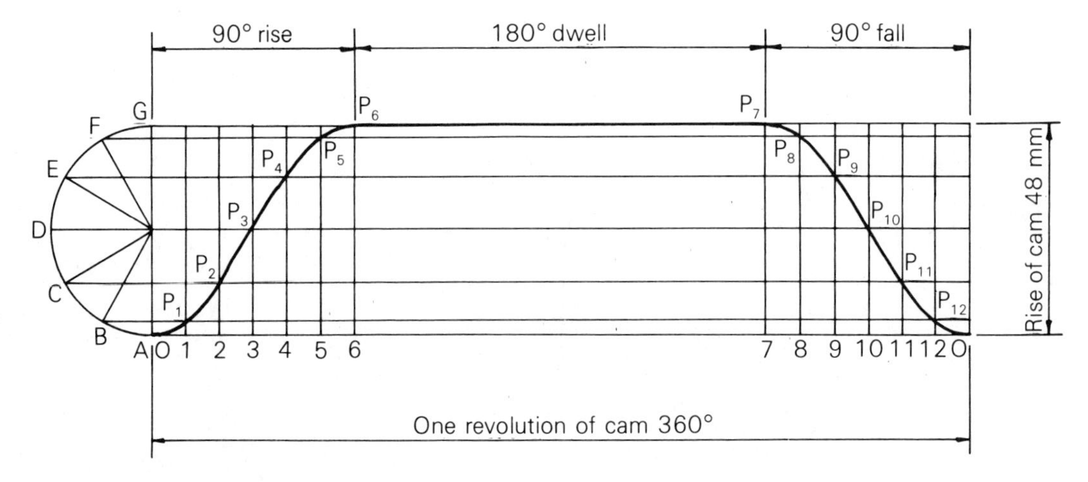

Fig. 2.25 Displacement diagram.

(b) To construct the cam profile (Fig. 2.26), draw a cylinder 60 mm diameter to represent the cam diameter.

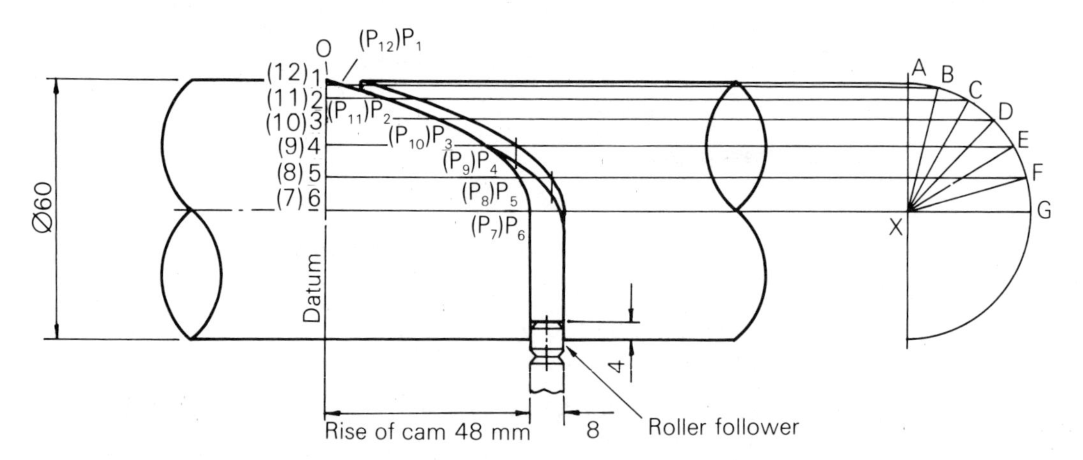

Fig. 2.26 Cylindrical cam profile.

(c) Draw an end view of the cylinder marking the radial positions of the vertical and horizontal axes A and G respectively. Angle AXG represents the rotation through which the cam rises (90°). Divide angle AXG into six equal parts and label as shown. Draw horizontal construction lines from each of these positions to the datum line on the cylinder and mark the points 0, 1, 2, 3, etc.

(d) On the displacement diagram, set dividers at the length of line $1P_1$ and transfer to the cam profile drawing to locate the position of P_1 from point 1 on the datum line.

(e) Repeat the transfer of distances from the displacement diagram to obtain positions P_2, P_3, P_4, etc., on the cam profile drawing.

(f) Draw a smooth curve through points O, P_1, P_2, etc., to complete one side of the cam profile for the follower rise.

(g) To complete the other side of the groove, locate positions displaced from points O, P_1, P_2, etc., by the amount equal to the roller width (8 mm). Draw a smooth curve through these positions to complete the groove for the cam profile for the follower rise.

(h) The dwell position will be represented on the profile drawing by a groove, 8 mm wide, parallel to the datum line. This completes one-half of the cam rotation.

(i) The depth of the slot is produced by dropping vertical lines from the intersections of the cam outer edges and angular displacement lines, equal in length to the depth of the groove (4 mm).

The student must appreciate that only half of the cam profile can be drawn in a single view in orthographic projection. It can be seen from the displacement diagram (Fig. 2.25) that the profile for the second half of rotation is a mirror image of that for the first half and consequently the second half of the cam profile will be a mirror image of the first half. The positions P_7, P_8, P_9, etc., shown in brackets in Fig. 2.26, are the relative positions in the second half of the cam profile.

In practice the groove width would be slightly larger than the roller diameter to allow for roller movement.

SELF-TEST 2

Select the correct option(s). *Note*: There may be more than one correct answer.

1. The smoothest follower motion is obtained using:

(a) an in-line knife-edge follower
(b) an offset knife-edge follower
(c) an in-line roller follower
(d) an offset roller follower.

2. The displacement diagrams shown in Fig. 2.27 describe

I uniform acceleration; II uniform velocity:

(a) I only
(b) II only
(c) both I and II
(d) neither I nor II.

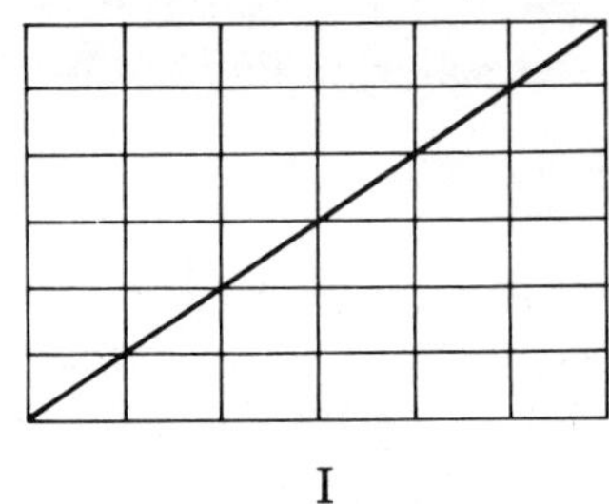

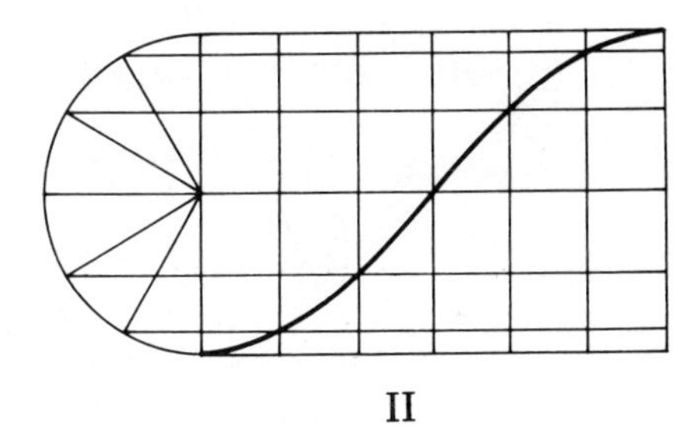

Fig. 2.27

3. A cylindrical cam is one where the follower movement is:

(a) at right angles to the cam axis
(b) parallel to the cam axis
(c) perpendicular to the axis of rotation
(d) parallel to the axis of rotation.

4. A flat follower is often mounted off-centre to:

(a) give a smoother action
(b) minimise any cam variation
(c) reduce wear
(d) reduce play in guides.

5. One of the disadvantages of uniform velocity motion is that it:

(a) is only suitable for high speeds
(b) can only be used on disc cams
(c) may provoke shock
(d) needs a dwell period.

6. Simple harmonic motion is preferred to uniform acceleration because it:

(a) provides a smoother action
(b) is cheaper to produce
(c) gives longer cam life
(d) does not require an offset follower.

7. The displacement diagram shown in Fig. 2.28 describes:

(a) rise 60 mm with uniform acceleration in 180°, instant fall 30 mm, fall 30 mm with uniform velocity in 180°
(b) rise 60 mm with simple harmonic motion in 180°, instant fall 30 mm, fall 30 mm with uniform velocity in 180°
(c) rise 60 mm with simple harmonic motion in 180°, dwell for 30 mm, fall 30 mm with uniform retardation in 180°
(d) rise 60 mm with uniform acceleration in 180°, dwell for 30 mm, fall 30 mm with uniform velocity in 180°.

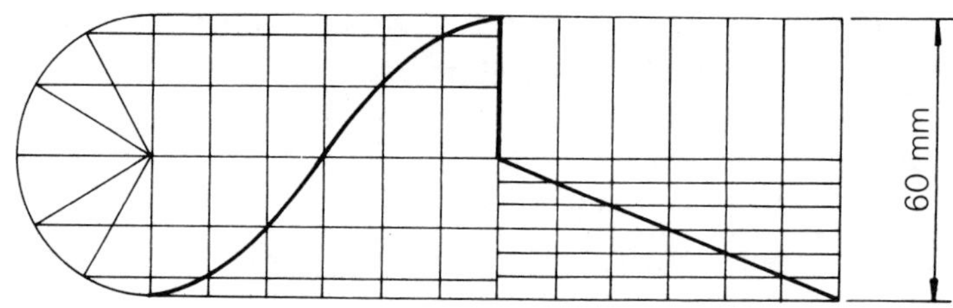

Fig. 2.28

8. The types of follower shown in Fig. 2.29 would be described respectively as:

(a) offset roller, in-line knife-edge, in-line roller, offset knife-edge
(b) offset roller, in-line knife-edge, in-line roller, in-line knife-edge
(c) in-line roller, offset knife-edge, offset roller, offset knife-edge
(d) offset roller, offset knife-edge, offset roller, in-line knife-edge.

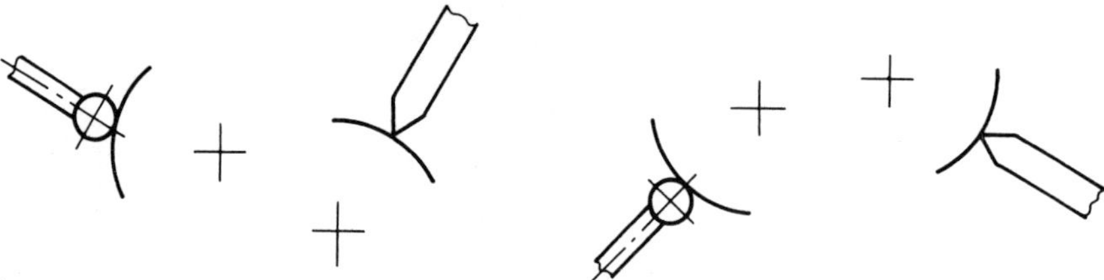

Fig. 2.29

9. The displacement diagrams shown in Fig. 2.30 describe

I uniform acceleration; II simple harmonic motion:

(a) I only
(b) II only
(c) both I and II
(d) neither I nor II.

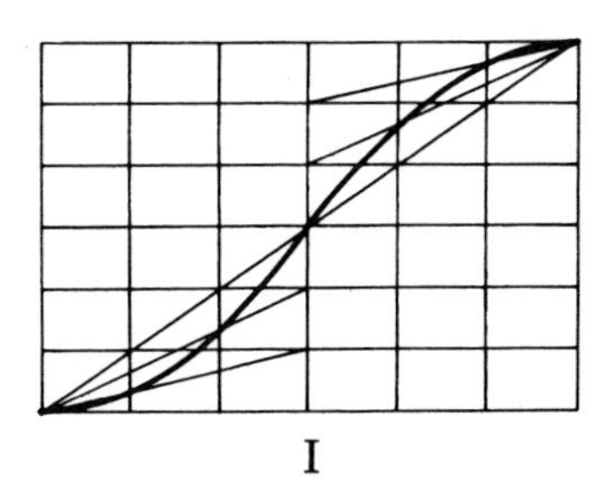

II

Fig. 2.30

10. Fig. 2.31 shows:

(a) one disc cam
(b) one cylindrical cam
(c) two disc cams
(d) two cylindrical cams.

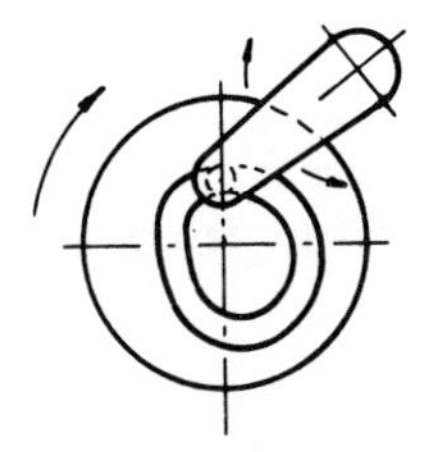

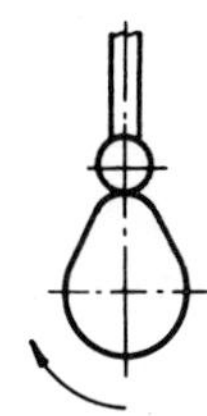

Fig. 2.31

EXERCISE 2

1. Sketch and describe a disc cam. What are the advantages and disadvantages of this type of cam?

2. Sketch three types of cam follower. What are the limitations in the practical application of each type?

3. What is the advantage in using uniform acceleration motion as compared to using uniform velocity?

4. Sketch and describe a cylindrical cam. What are the advantages and disadvantages of this type of cam?

5. Three types of motion imparted to a follower are: uniform velocity, uniform acceleration and simple harmonic motion. Give a practical example of the use of each of these motions.

6. What are the advantages and dangers in offsetting:

(a) flat, (b) knife-edge and (c) roller followers?

7. What is a displacement diagram? Why is a displacement diagram drawn while preparing a cam profile drawing for a given specification?

8. During the construction of a cam profile drawing a series of points is usually plotted. Why are these points joined by a smooth curve instead of a series of straight lines?

9. Construct a displacement diagram for a disc cam to the following specification:

minimum radius of cam 45 mm
follower to:
rise 35 mm with uniform acceleration in 120° of cam rotation
dwell for 60° of cam rotation
fall 35 mm with uniform retardation for the remaining 180° of cam rotation
rotation of cam anti-clockwise.

10. Construct a cam profile for a disc cam with an in-line knife-edge follower to the specification listed in Question 9.

11. Construct a cam profile for a disc cam with an in-line roller follower, 20 mm diameter, to the specification listed in Question 9.

12. Construct a cam profile for a disc cam with an offset roller follower, 20 mm diameter, to the following specification:

amount of offset 10 mm
minimum radius of cam 35 mm
follower to:
rise 20 mm with uniform velocity in 180° of cam rotation
dwell for 90° of cam rotation
fall 20 mm with uniform velocity in 90° of cam rotation
rotation of cam clockwise.

13. Construct a cam profile for a disc cam with an offset roller follower, 25 mm diameter, to the following specification:

amount of offset 15 mm
minimum radius of cam 40 mm
follower to:
rise 25 mm with simple harmonic motion in 90° of cam rotation
dwell for 45° of cam rotation
fall 25 mm with simple harmonic motion in 180° of cam rotation
dwell for remaining 45° of cam rotation
rotation of cam anti-clockwise.

14. Construct a cam profile for a disc cam with a radial arm roller follower which pivots about a point 90 mm to the left and in line with the horizontal axis of the cam. The specification is as follows:

length of pivot arm 80 mm (to roller centre)
roller follower 20 mm diameter
minimum radius of cam 80 mm
follower to:
rise 20 mm with simple harmonic motion in 180° of cam rotation
drop 5 mm instantly and dwell for 90° of cam rotation
fall 15 mm with simple harmonic motion in 60° of cam rotation
dwell for the remaining 30° of cam rotation
rotation of cam clockwise.

15. Construct a cam profile for a cylindrical cam to the following specification:

diameter of cam 50 mm
diameter of roller follower 10 mm
depth of slot 5 mm
follower to:

rise 60 mm with simple harmonic motion in 120° of cam rotation
dwell for 60° of cam rotation
fall 60 mm with simple harmonic motion in 120° of cam rotation
dwell for 60° of cam rotation.

SECTION C

SCREW THREADS

The Design and Use of Commercially Available Screw Threads

After reading this chapter you should be able to:

★ appreciate the design and use of commercially available screw threads as specified in British Standards (G);

★ recognise the basic forms of standard screw thread (S);

★ sketch and label standard screw threads (S);

★ list applications relating to standard thread forms (S);

★ construct helices with reference to single and multi-start threads (S).

(G) = general TEC objective

(S) = specific TEC objective

INTRODUCTION

The use of screws in a general engineering capacity has been known for over 2000 years, although they were seldom used until the later Middle Ages. As Great Britain developed into an industrial nation, the need for standardisation and interchangeability became readily apparent, and in 1841 Sir Joseph Whitworth attempted to establish a uniform screw thread standard based on a 55° vee-form.

About the same time a 60° vee-form thread came into general use in the United States which ultimately led to the American National form and was the basis for the unified screw threads agreed by the United States, Canada and the United Kingdom as a standard to obtain interchangeability between these three nations.

With the advent of the Second World War the need for world-wide standardisation and interchangeability was even more evident, and it ultimately led to the establishment of two international (ISO) standards for unified and metric screw threads. The latter has now been strongly recommended for adoption by UK industry with the former as second choice.

SCREW THREAD TERMINOLOGY

A *screw thread* is the ridge produced by forming a continuous helical groove on the surface of a cylinder.

It may be termed an *external* (*male*) *screw thread*, when formed on the external surface of a cylinder, for example, as the thread on a bolt, or an *internal* (*female*) *screw thread*, when formed on the internal surface of a hollow cylinder, for example, as the thread in a nut—see Fig. 3.1).

A *right-hand screw thread* is a thread which, if assembled with a stationary mating thread, moves away from the observer when rotated in a clockwise direction. A *left-hand screw thread*, when assembled in a similar way, moves away from the observer when rotated in an anti-clockwise direction. A thread formed by a single continuous

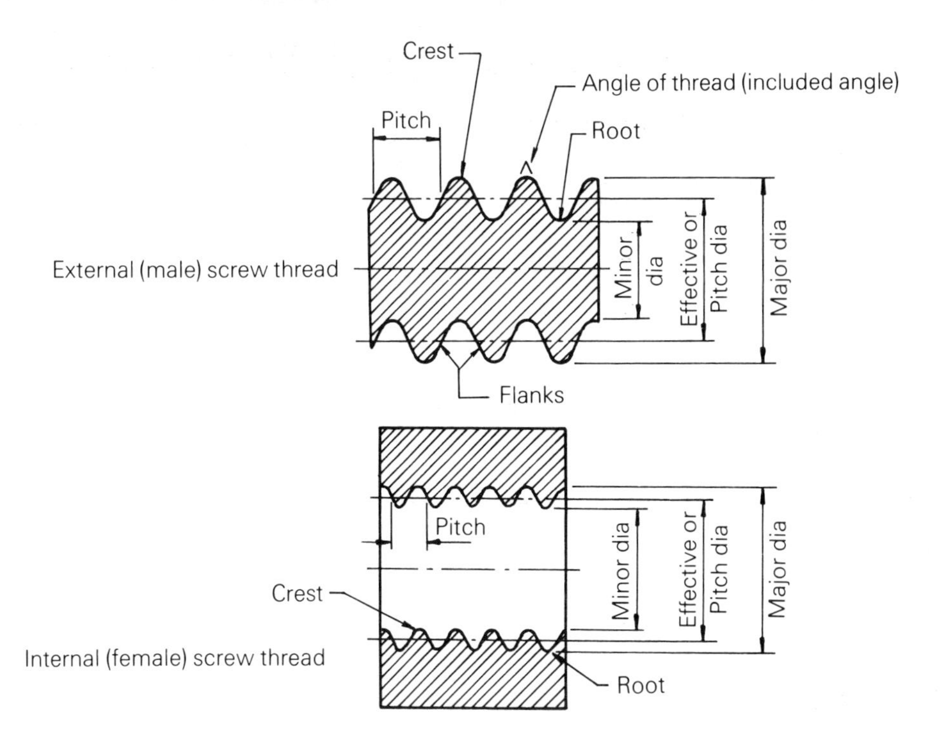

Fig. 3.1 Screw thread terminology.

helical groove is called a *single-start screw thread* and a thread formed by a combination of two or more helical grooves equally spaced along the axis is called a *multi-start screw thread*—see Fig. 3.2.

The *flanks* of a thread are those parts of the surface on either side of the thread.

The *pressure flank* is the flank which takes a load in an assembly.

The *clearing flank* is the flank which does not take any load in an assembly.

The *crest* is that part of the surface of a thread which connects adjacent flanks at the top of the ridge.

The *root* is that part of the surface which connects to adjacent flanks at the bottom of the groove.

The *angle of thread* is the included angle between the flanks—see Fig. 3.1.

The *pitch* is the distance, measured parallel to the axis, between corresponding points on adjacent thread forms (Figs 3.1 and 3.2).

The *lead* is the distance, measured parallel to the axis, between corresponding points on consecutive contours of the same thread helix (Fig. 3.2). The lead is the distance the thread advances axially in one revolution. For a single-start thread the lead is identical to the pitch. For a two-start thread the lead = 2 × pitch, and so on.

The *major diameter* is the diameter over the crests of an external thread and between the roots of an internal thread.

The *minor diameter* is the diameter between the roots of an external thread and between the crests of an internal thread.

The *effective (or pitch) diameter* is the diameter of an imaginary cylinder where the widths of the thread and the groove are equal (see Fig. 3.1).

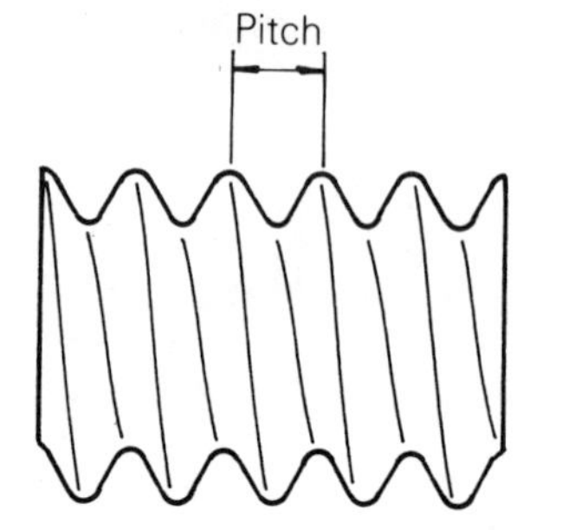

Single-start right-hand screw thread

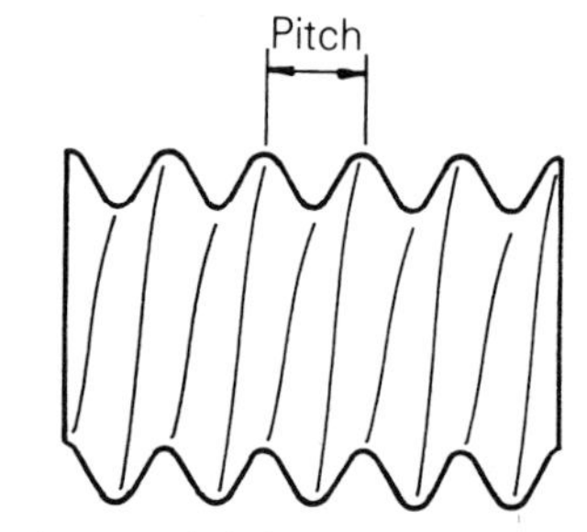

Single-start left-hand screw thread

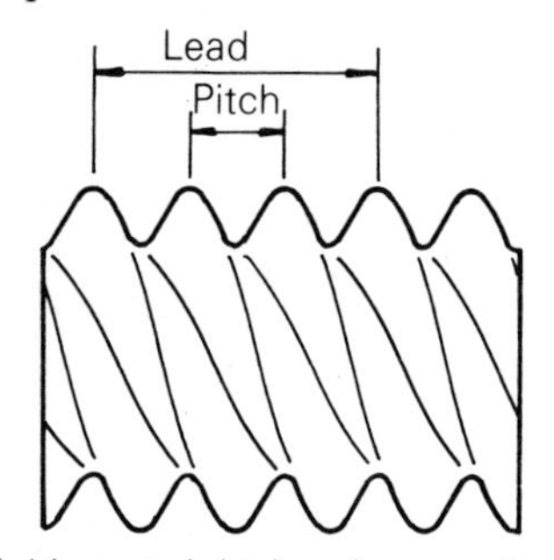

Multi-start right-hand screw thread

Fig. 3.2 Single and multi-start screw threads.

COMMERCIALLY AVAILABLE SCREW THREADS

(a) Vee-form Threads

(i) ISO metric screw threads, BS 3643: Part 1: 1963

The basic profile of ISO metric threads (Fig. 3.3) is a symmetrical vee-thread which has an included angle of 60°.

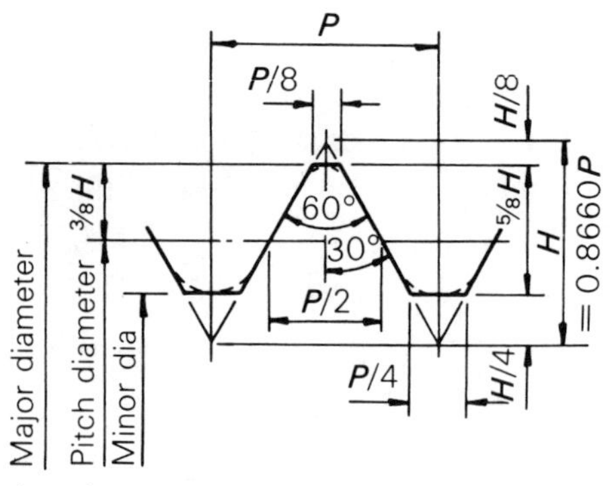

Fig. 3.3 ISO metric screw thread.

The crests of bolts and the roots of nuts are basically flat but in practice they are generally rounded as shown.

The Standard specifies two series of diameters with graded pitches, one with coarse and the other with fine pitches. The basic major diameters for the recommended series for screws, bolts and nuts range from 1 mm to 300 mm.

Screw threads to this Standard are designated by the letter M followed by the value of the nominal diameter and the pitch, both in millimetres, for example:

M8 × 1.25

The absence of an indication of pitch means that a coarse thread is specified, for example, a coarse thread M20 × 2.5 may be designated M20.

(ii) Unified screw threads, BS 1580: Parts 1 and 2: 1962

The basic profile is identical to the metric screw thread, having a symmetrical vee-thread of included angle 60° (Fig. 3.3).

The unified system is recommended by the International Organization for Standardization (ISO) as the international system of screw threads in inch units, in parallel with the similar international system of screw threads in metric units (BS 3643).

Screw threads to the unified Standard are designated by specifying in sequence the nominal size, number of threads per inch, and thread series symbol (for example, UNC for coarse threads, UNF for fine threads and UNEF for extra-fine threads). For example:

1/4 —20 UNC
0.250 —20 UNC
9/16 —20 UNF
0.5625—20 UNF

The standard series unified screw threads in BS 1580: Parts 1 and 2 relate to the ranges:

coarse thread, UNC	$\frac{1}{4}$ to 4 inches diameter
fine thread, UNF	$\frac{1}{4}$ to $1\frac{1}{2}$ inches diameter
extra fine thread, UNEF	$\frac{1}{4}$ to $1\frac{11}{16}$ inches diameter

Part 3 of BS 1580 relates to diameters less than $\frac{1}{4}$ inch and it will be noted that the range of diameters falls within the range of the obsolete BS 93, British Association (BA) screw threads, which is detailed later.

BS 1580: Part 3 relates to unified screw threads from Number 0 (0.060 inch diameter) to Number 12 (0.216 inch diameter), commonly referred to as the 'numbered sizes'. The recommended method of designating these sizes is to specify in sequence the designation number followed (if desired) by the nominal size in brackets, then the number of threads per inch and the thread series symbol. For example:

0(0.060)—80 UNF
2(0.086)—56 UNC
12(0.216)—32 UNEF

The following threads are still commercially available but are not used in the construction of new equipment. They will undoubtedly continue to be used, however, for many years, especially on replacements or spare parts.

(iii) Screw threads of Whitworth form, BS 84: 1956

This British Standard is now obsolete. It includes:

(1) a coarse thread series, the British Standard Whitworth (BSW) Series, from $\frac{1}{8}$ inch to 6 inch diameter
(2) a fine thread series, the British Standard Fine (BSF) Series, from $\frac{3}{16}$ inch to $4\frac{1}{4}$ inch diameter.

The basic form of the Whitworth thread (Fig. 3.4) is a symmetrical vee-thread which has an included angle of 55°.

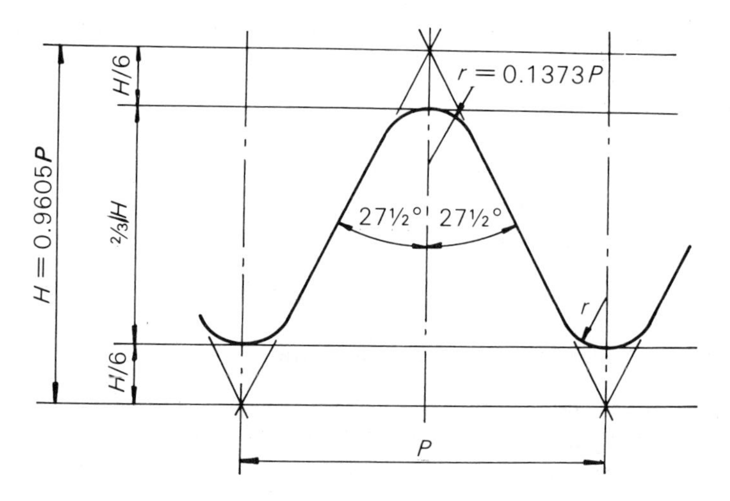

Fig. 3.4 Basic form of Whitworth thread.

Screw threads to this Standard are designated by the major diameter of the thread and the number of threads per inch followed by the thread series symbol. For example:

1 in − 8 BSW
$\frac{1}{2}$ in − 16 BSF

(iv) British Association (BA) screw threads, BS 93: 1951

This British Standard is also obsolete. It relates to the basic form, as shown in Fig. 3.5, and is a symmetrical vee-thread which has an included angle of $47\frac{1}{2}$°.

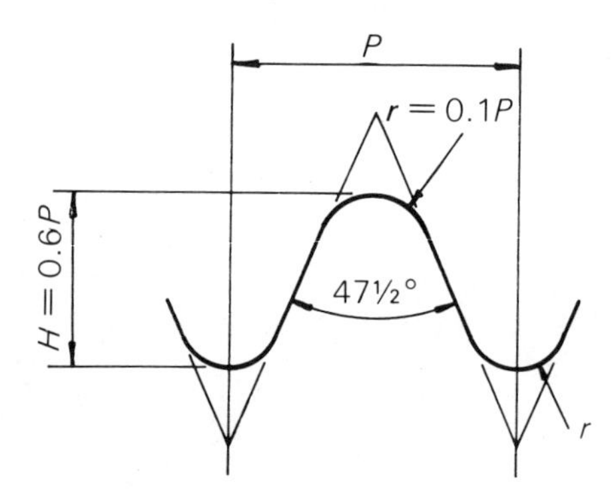

Fig. 3.5 BA (British Association) thread.

BA threads are a numbered range covering 26 sizes from 0 BA (6.00 mm) to 25 BA (0.25 mm).

Screw threads to this Standard are designated by the series number followed by the thread series symbol. For example:

2 BA

One Whitworth form thread which has been kept in the metric plan is the pipe thread.

(v) Pipe threads, BS 21: 1973 Pipe threads where pressure-type joints are made on threads; BS 2779: 1973 Pipe threads where pressure-type joints are not made on threads

These Standards have replaced the earlier editions which designated pipe threads as 'British Standard Pipe (BSP) threads'.

Pipe threads are of Whitworth form for pipe sizes $\frac{1}{16}$ inch to 6 inch bore.

BS 21 relates to taper external threads for assembly with either taper or parallel internal threads (the latter being designated as Rp$\frac{1}{2}$ for $\frac{1}{2}$ inch pipes) and BS 2779 relates to threads for fastening purposes, such as the mechanical assembly of the component parts of fittings, cocks and valves, etc., which are designated by the letter G followed by the pipe size, for example, G$\frac{1}{2}$ for $\frac{1}{2}$ inch pipes.

Dimensions of the thread are given in millimetres (compatible inch sizes also being given).

(b) Threads to Transmit Power

Vee-form threads are not particularly suitable for transmitting power. More acceptable thread forms are available for this purpose.

(i) Square threads

The square thread (Fig. 3.6) is difficult to produce economically because of the straight sides. It is sometimes modified by providing a slight taper to the sides but has been generally replaced by the acme and trapezoidal screw threads.

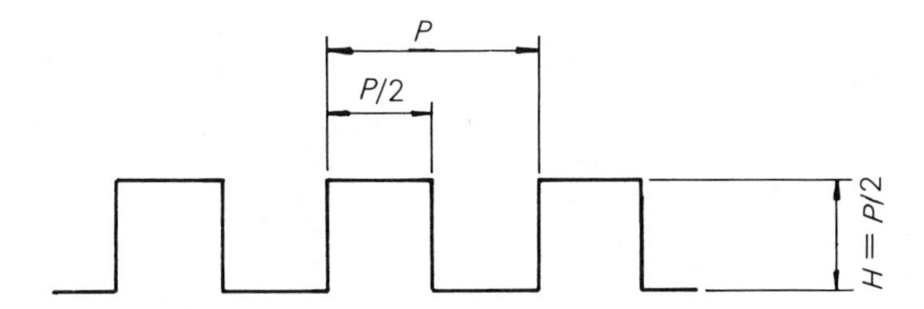

Fig. 3.6 Square thread.

(ii) Acme screw threads, BS 1104: 1957

The acme thread (Fig. 3.7) has generally replaced the square thread. It is produced more easily, is stronger and allows the use of a split nut (particularly useful in machine tool lead screw applications).

Screw threads to this Standard are designated by the major diameter (in inches) of the thread, followed by the number of threads per inch and then the name of the thread. For example:

$1\frac{1}{2}-4$ ACME

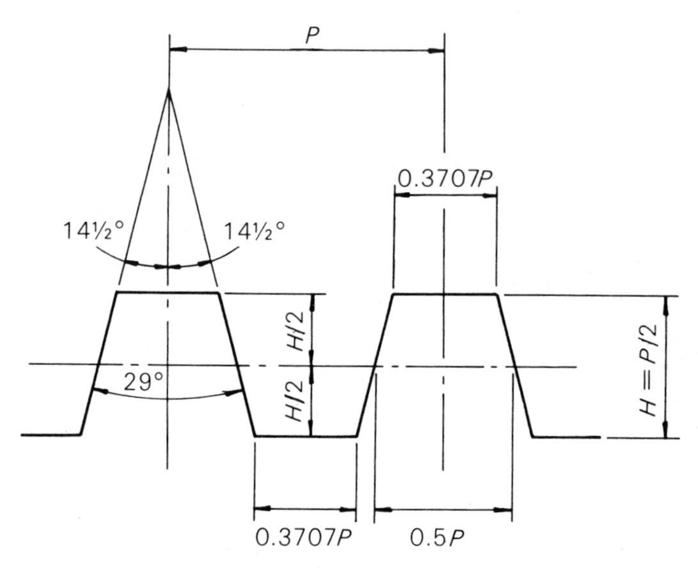

Fig. 3.7 Acme thread.

(iii) ISO metric trapezoidal screw threads, BS 5346: 1976

The trapezoidal screw thread (Fig. 3.8) is very similar to the acme thread and will continue to replace it in the construction of new equipment.

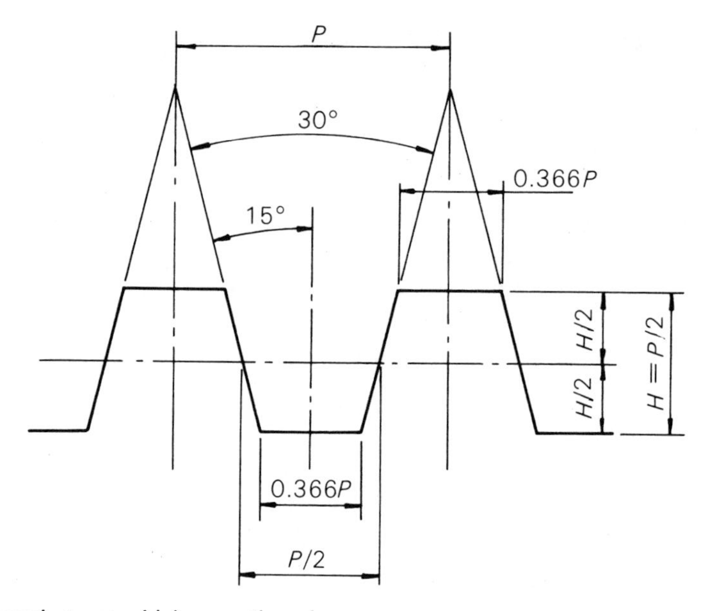

Fig. 3.8 ISO metric trapezoidal screw thread.

A screw thread conforming to this standard is designated by the letters Tr followed by the major diameter and the pitch in millimetres and separated by the sign ×. For example:

Tr 20 × 4

(iv) Buttress threads, BS 1657: 1950

The buttress thread is used to transmit power in one direction only. The standard form has a pressure flank angle of 7° and an included angle of 52° to the clearing flank—see Fig. 3.9.

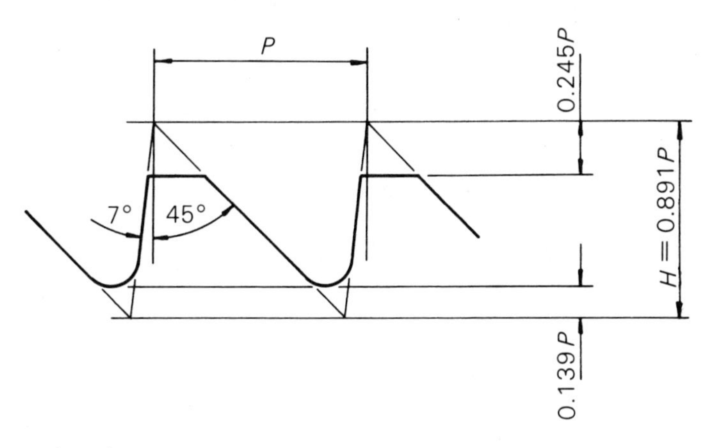

Fig. 3.9 Buttress thread.

Screw threads to this Standard are designated by the major diameter (in inches) of the thread followed by the name 'BS Buttress thread' and then the number of threads per inch. For example:

2.0 BS Buttress 8 t.p.i.

SKETCHING THE BASIC FORM OF STANDARD SCREW THREADS–

(i) ISO Metric Screw Thread

Problem: To construct the form of an external (male) thread M36 × 4 to a scale of 10:1.

Solution: See Fig. 3.10.

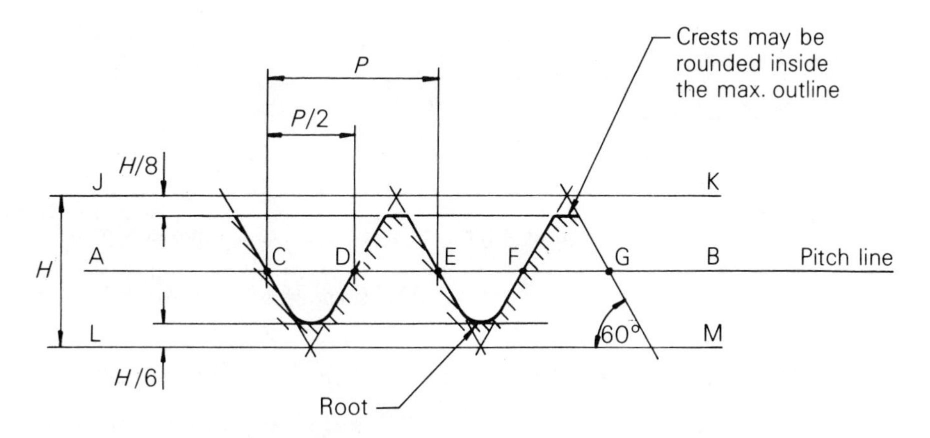

Fig. 3.10 Construction of ISO metric screw thread.

(a) Construct horizontal line AB to represent the pitch line of the thread.

(b) Along the pitch line mark off points C and D to represent the half-pitch, $P/2$, of the thread (in this instance 2 mm × scale ratio = 20 mm).

(c) Similarly mark off points E, F and G and draw in lines, at 60° to the horizontal, from each of these points.

(d) Join the apex points to give lines JK and LM, which are distance H apart.

(e) Draw lines $H/8$ from JK and $H/6$ from LM to give the positions of the crest and root of the thread.

(f) Using compasses (radius curves facilitate this task) draw in the root arcs of the thread.

(g) Complete the thread profile by drawing in the thick outlines.

(ii) ISO Metric Trapezoidal Screw Threads

Problem: To construct the form of an external (male) thread Tr 220 × 20 to a scale of 5:1.

Solution: See Fig. 3.11.

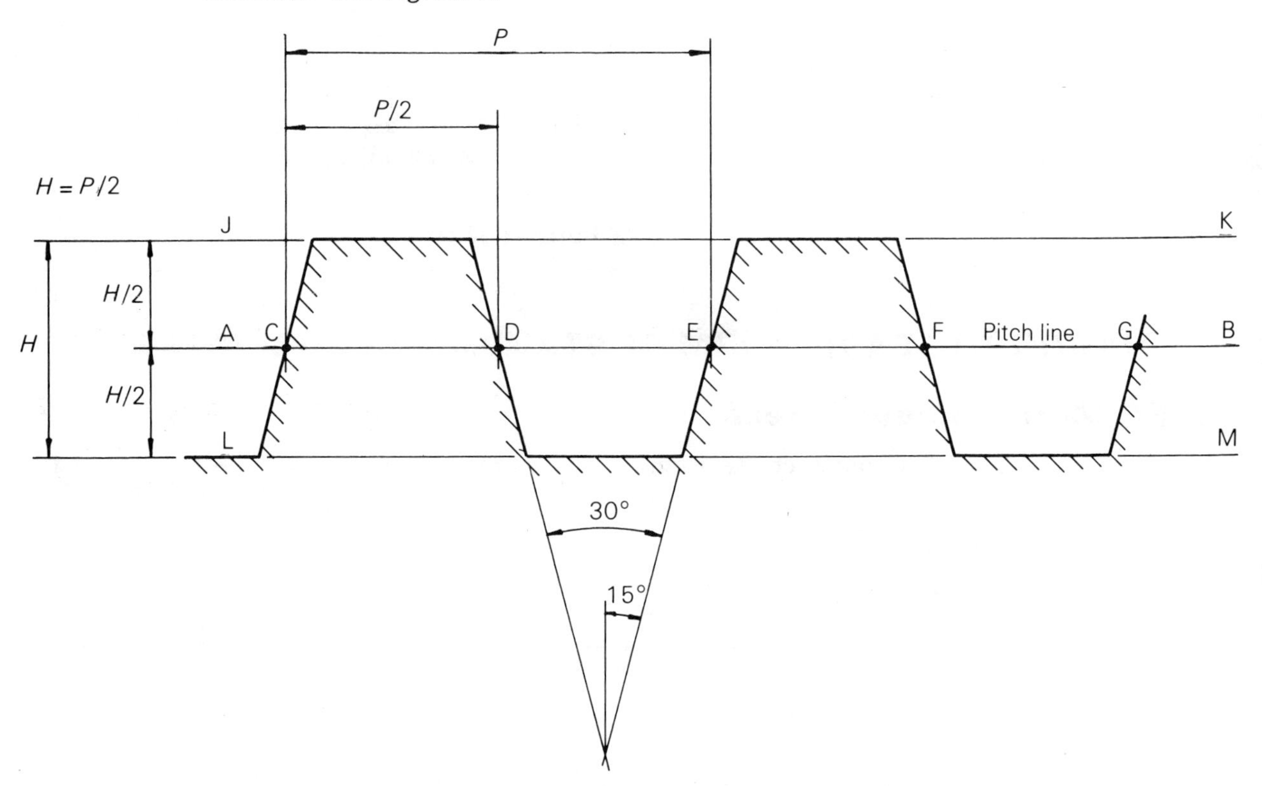

Fig. 3.11 Construction of ISO metric trapezoidal screw thread.

(a) Construct a horizontal line AB to represent the pitch line of the thread.

(b) Along the pitch line mark off points C and D to represent the half-pitch, $P/2$, of the thread (in this instance 10 mm × scale ratio = 50 mm).

(c) Similarly mark off points E, F and G and draw in lines, inclined at 15° to the vertical, from each of these points to meet lines JK and LM, which are each $H/2$ distant from the pitch line. (*Note*: in a trapezoidal thread $H = P/2$.)

(d) Complete the thread profile by drawing in the thick outlines.

(iii) Buttress Thread

Problem: To construct the form of a buttress thread with 2.5 t.p.i., to a scale of 5:1.

Solution: See Fig. 3.12.

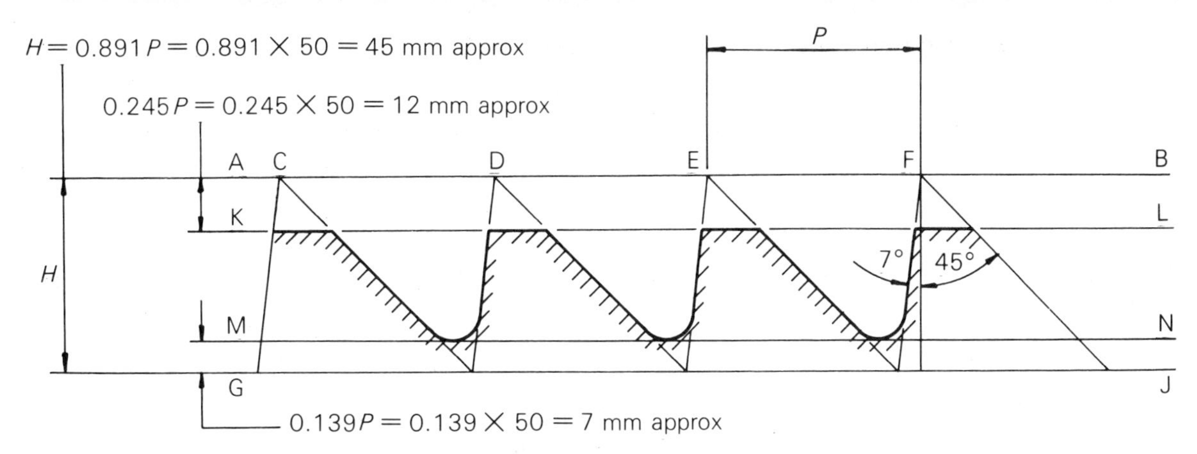

Fig. 3.12 Construction of a buttress thread.

(a) Draw a horizontal line AB and mark off points C and D at a distance equal to the pitch, P, 50 mm approx (for 2.5 t.p.i., pitch $= \frac{1}{2.5} = 0.4$ inch at scale ratio 5:1, pitch = 2 inch, i.e. 50 mm approx).

(b) Similarly mark off points E and F and draw in lines, inclined at 7° and 45° to the vertical, to represent the pressure and clearing flanks respectively. The intersections of these lines meet at line GJ, a distance H from line AB. (*Note*: for a buttress thread $H = 0.891P$.)

(c) Draw in lines KL, at a distance $0.245P$ from line AB, and MN, a distance $0.139P$ from GJ, to locate the crests and roots of the thread.

(d) Using compasses (radius curves facilitate this task) draw in the root arcs of the thread.

(e) Complete the thread profile by drawing in the thick outlines.

APPLICATIONS RELATING TO STANDARD THREAD FORMS

The applications relating to standard vee-form threads are listed in Table 3.1 and those relating to power transmitting threads in Table 3.2.

Table 3.1 *Applications for vee-form threads*

Thread form		*Applications*
Coarse threads	metric coarse UNC BSW (obsolete)	Used for general purposes — fasteners, particularly useful on low-tensile materials such as cast iron, aluminium, magnesium and plastics to reduce the risk of thread stripping
Fine threads	metric fine UNF BSF (obsolete)	Fine threads have greater tensile stress area than coarse threads. Particularly used where length of engagement is short or for thin wall thickness. Can withstand the loosening forces of vibration better than coarse threads
Pipe threads		Whitworth form with fine pitch. Used for tubes, pipes and fittings
BA (obsolete)		Used in preference to Whitworth form for sizes <¼ inch. Applications include use in instrument industries, meters, and electrical and radio fittings

Vee-form threads are used for general fastener applications (nuts, screws, bolts), for adjusting purposes and in measuring instruments.

Power-transmitting threads are subject to less frictional force than vee-form threads due to the pressure flank being nearer perpendicular to the line of thrust, thus reducing oblique outward forces.

Table 3.2 *Applications for power-transmitting threads*

Thread form	*Applications*
Square Acme Metric trapezoidal	Fly presses, screw jacks, valve spindles, lead screws. Square threads are difficult to manufacture because of the parallel sides of the profile. Relatively weak at the root because the width is only half the pitch. Acme threads, made in imperial sizes, replaced square threads for the above applications. With the advent of metrication the acme thread is being superseded by the metric trapezoidal thread
Buttress	Used when heavy forces are applied in one direction only, such as in quick release vice screws. Also used for screwing together tubular members, for example, breech mechanisms of large guns

THE HELIX

The helix is a curve generated by a point moving uniformly along the surface of a cylinder while the cylinder is rotating uniformly about its axis—see Fig. 3.13.

A screw thread is an example of the application of helix construction and may be 'right-hand' or 'left-hand' (see Screw Thread Terminology, p. 48).

Example 1 Construct the profile for a single-start right-hand ISO metric trapezoidal screw thread Tr 140 × 24, scale 1:1.

Procedure: From BS 5346, Metric trapezoidal screw threads, the following information is confirmed:

Major diameter 140 mm
Minor diameter 116 mm
Pitch 24 mm
Width of thread at major diameter $= 0.366P = 0.366 \times 24 = 8.784$
(for drawing purposes, say 9 mm)
Included angle 30°

Construction: See Fig. 3.14.

(a) Construct two semi-circles, of diameters 140 and 116 mm, to represent the major and minor diameters of the thread in half end-view. Divide these into six equal parts and number as shown.

(b) Draw horizontal lines from each of these points across the elevation of the thread.

(c) Mark off, say, four pitches, each 24 mm long, and divide each pitch into twelve equal parts and number as shown.

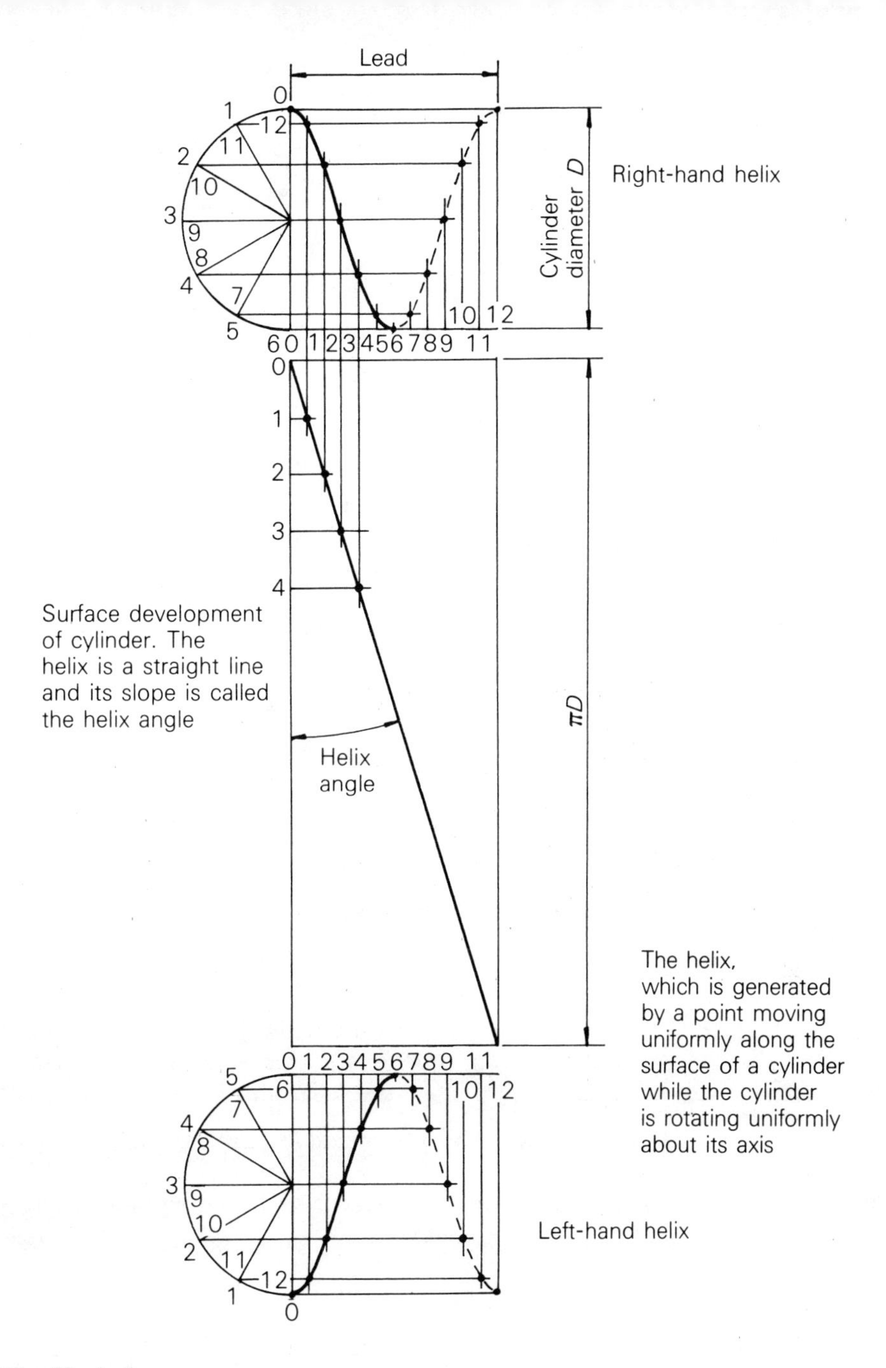

Fig. 3.13 The helix.

(d) To construct the helix of the major diameter locate point 1 on the major diameter semi-circle and follow along the horizontal line until it crosses the vertical line 1. Mark this junction with a dot. Follow this procedure with each point in turn and join each dot with a smooth curve to produce the helix for a length of one pitch.

(e) Repeat this procedure to obtain the helix of the minor diameter, taking particular care to start from the correct position.

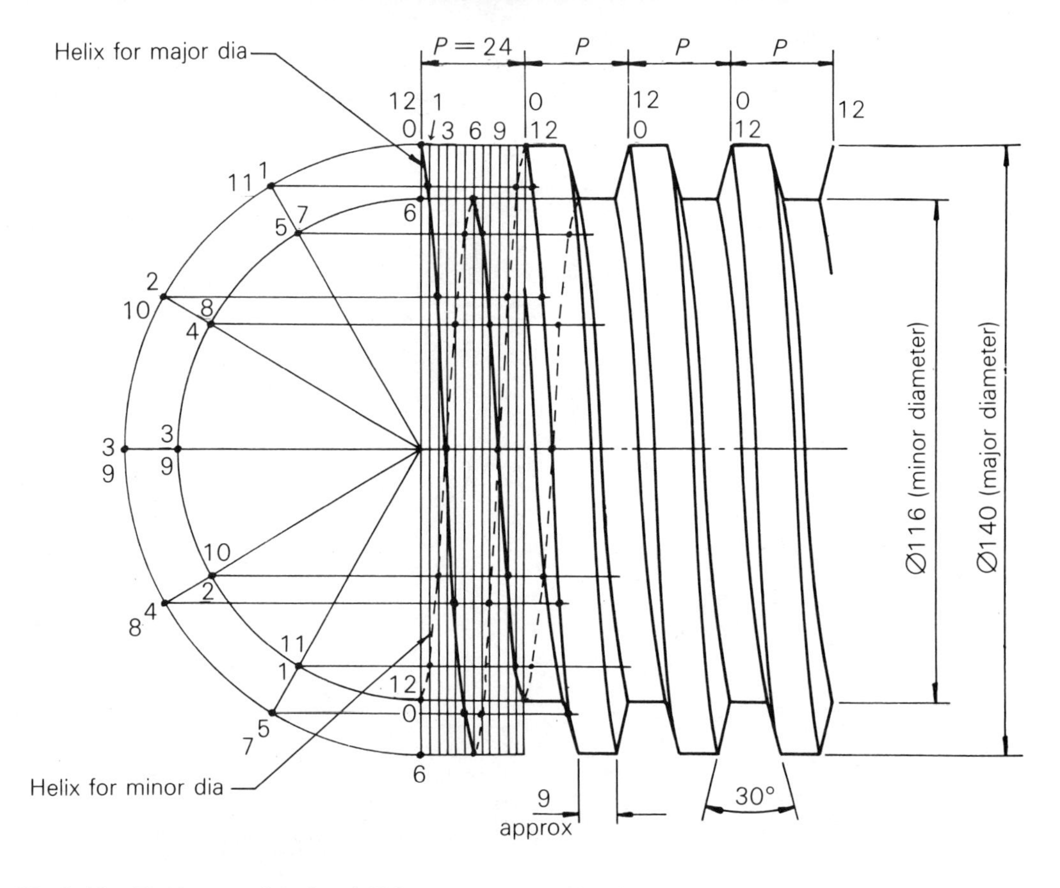

Fig. 3.14 Single-start right-hand ISO metric trapezoidal thread.

(f) Repeat steps (d) and (e) to obtain the helix for the remaining three pitches to construct the thread profile and draw in the included thread angle of 30° (vertical construction lines have been deleted from the illustration for clarity).

(g) Complete the elevation by drawing in visible outlines only.

Example 2 Construct the profile for a two-start right-hand square thread with major diameter 120 mm and lead 48 mm, scale 1:1.

Procedure: For a multi-start thread:

$$\text{pitch} = \frac{\text{lead}}{\text{number of starts}}$$

$$\therefore \quad \text{pitch, } P = \frac{48}{2} = 24\,\text{mm}$$

$$\text{width of thread} = \frac{P}{2} = 12\,\text{mm}$$

$$\text{depth of thread} = \frac{P}{2} = 12\,\text{mm}$$

$$\therefore \quad \text{minor diameter} = 96\,\text{mm}$$

Construction: See Fig. 3.15.

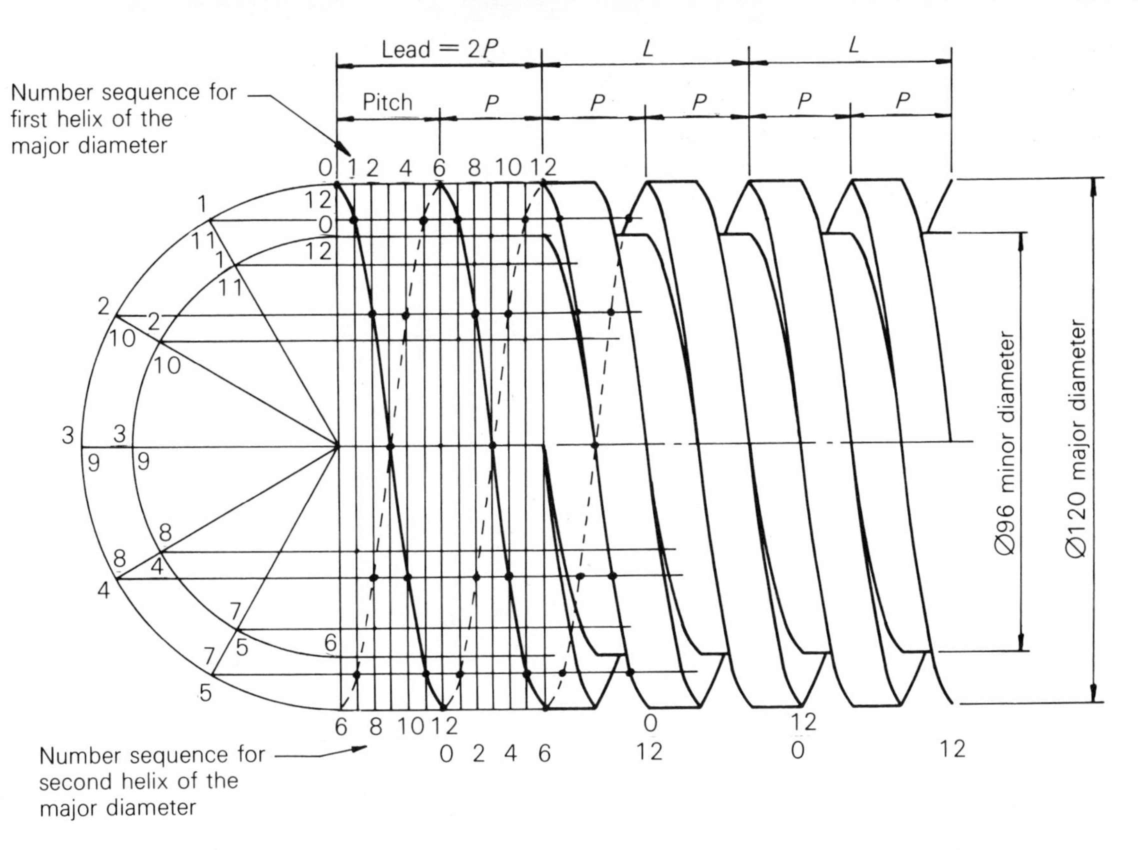

Fig. 3.15 Two-start right-hand square thread.

(a) Construct two semi-circles, of diameters 120 and 96 mm, to represent the major and minor diameters of the thread in half end-view. Divide these into six equal parts and number as shown.

(b) Draw horizontal lines from each of these points across the elevation of the thread.

(c) Mark off, say, 3 × the lead, each 48 mm long, and divide each lead into twelve equal parts and number as shown. (*Note*: in two-start systems, one helix is 180° in advance of the other and needs to be numbered in different sequence.)

(d) To construct the first helix of the major diameter locate point 1 on the major diameter semi-circle and follow along the horizontal line until it crosses the vertical line 1. Mark this junction with a dot. Follow this procedure with each point in turn and join each dot with a smooth curve to produce the visible helix for a length of one pitch.

(e) Repeat step (d) to construct the visible second helix of the major diameter starting from the junction of lines 1.

(f) Repeat steps (d) and (e) to obtain the helices of the minor diameter, taking particular care to start from the correct position (these helices have been deleted from the illustration for clarity).

(g) Repeat steps (d), (e) and (f) for the remaining leads to construct the thread profile (vertical construction lines have been deleted from the illustration for clarity).

(h) Complete the elevation by drawing in visible outlines only.

Example 3 Construct the profile of a single-start left-hand square thread with major diameter 60 mm and pitch 24 mm, scale 1:1.

Procedure: In Fig. 3.16a, for clarity, only two helices have been drawn. In Fig. 3.16b, the complete visible outline of the thread profile is constructed. Can you draw a similar profile without instructions?

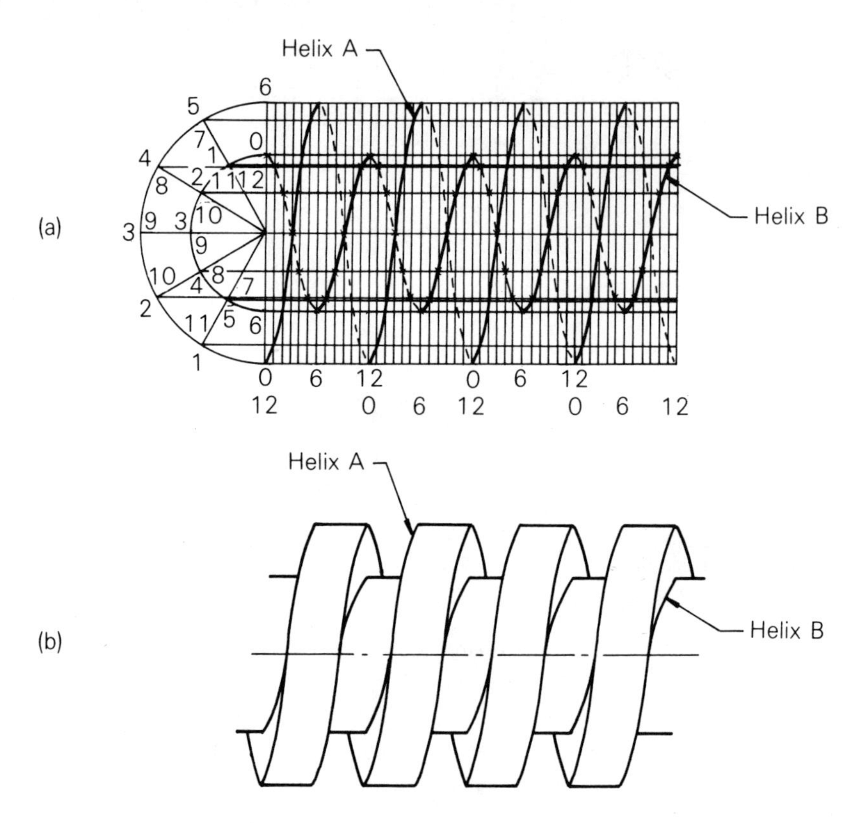

Fig. 3.16 Single-start left-hand square thread.

SELF-TEST 3

Select the correct option(s). *Note*: There may be more than one correct answer.

1. A nut which moves a distance of one pitch in 120° of rotation has the following number of starts:

(a) one
(b) two
(c) three
(d) four.

2. The major diameters of a female thread and a male thread are, respectively, the diameter:

(a) over the roots and between the crests
(b) between the roots and over the crests
(c) over the crests and between the roots
(d) between the crests and over the roots.

3. The angle of thread of an ISO metric screw is:

(a) 60°
(b) 55°
(c) $47\frac{1}{2}°$
(d) 30°.

4. The thread which has unequal pressure flank and clearance flank angles is the:

(a) buttress
(b) metric trapezoidal
(c) square
(d) acme.

5. The designation for an ISO metric trapezoidal screw thread of nominal diameter 60 mm and pitch 4 mm is:

(a) Tr M60 × 4
(b) Tr 4 × 60
(c) Tr M4 × 60
(d) Tr 60 × 4.

6. The power-transmitting thread used to accept load in one direction only is the:

(a) square thread
(b) buttress thread
(c) ISO metric trapezoidal thread
(d) acme thread.

7. The square thread shown in Fig. 3.17 is called a:

(a) two-start right-hand thread
(b) single-start right-hand thread
(c) two-start left-hand thread
(d) single-start left-hand thread.

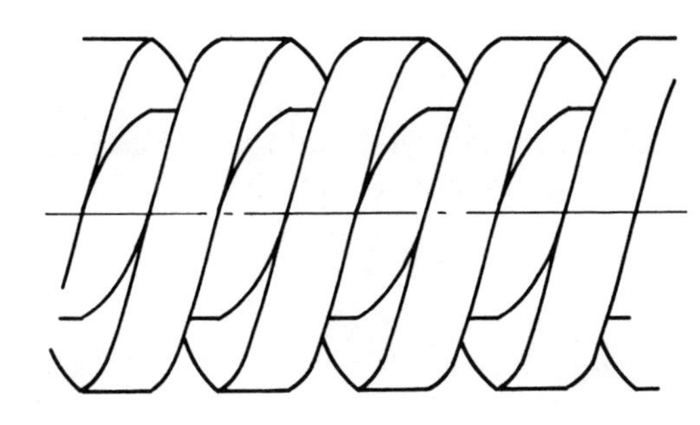

Fig. 3.17

8. The vee-form thread recommended for use on low-tensile materials, such as aluminium, is the:

(a) metric fine thread
(b) metric coarse thread
(c) pipe thread
(d) unified fine thread.

9. For medium-tensile stress conditions the most suitable thread to be adopted is the:

(a) UNF
(b) metric coarse
(c) UNC
(d) metric fine.

10. The lead of a two-start thread with 10 mm pitch is:

(a) 20 mm
(b) 10 mm
(c) 5 mm
(d) $\frac{1}{5}$ mm.

EXERCISE 3

1. Name the types of vee-form thread shown in Fig. 3.18.

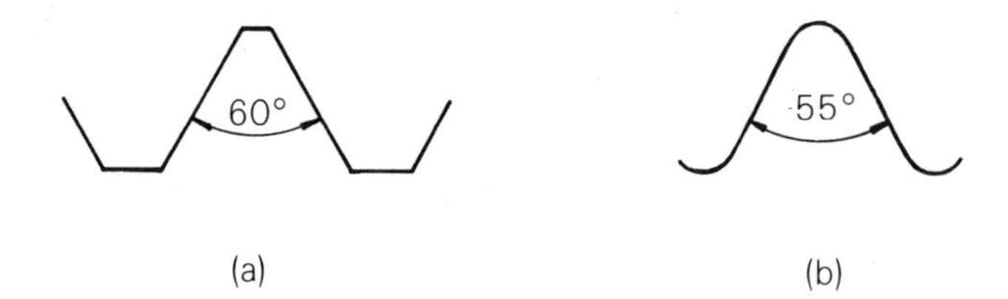

Fig. 3.18

2. Name the types of power-transmitting thread shown in Fig. 3.19.

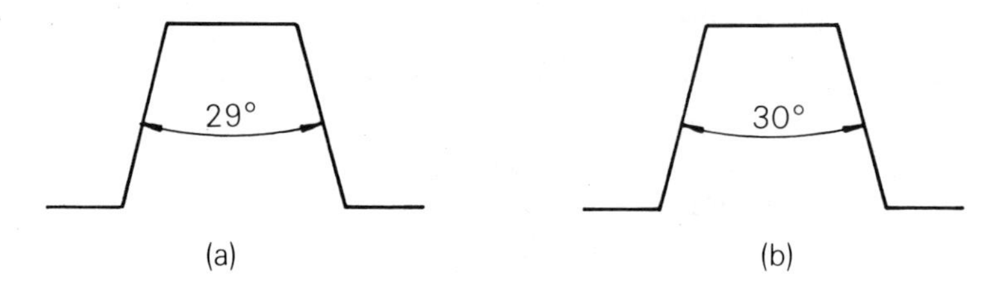

Fig. 3.19

3. Fig. 3.20 shows the sectional view of a male thread. What is the terminology used to describe the parts labelled a to h?

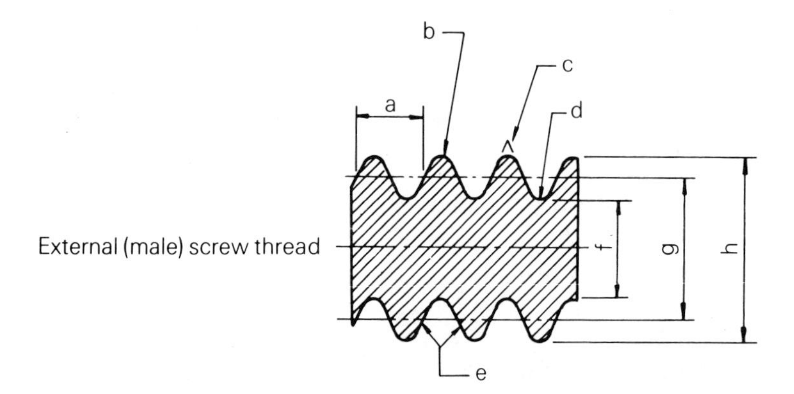

Fig. 3.20

4. How many starts has the thread of a nut which moves axially a distance of 75 mm while making five revolutions if the pitch is 5 mm?

5. Sketch two pitches of a Whitworth form thread, of pitch 15 mm at scale 5:1. Mark on your sketch the angle of thread and the pitch.

6. Construct a right-hand helix which would be generated by a point moving along a cylinder of diameter 75 mm during four revolutions if the lead of the helix is 48 mm.

7. Construct two pitches of the form of an external thread M48 × 5 to a scale of 10:1.

8. Construct the profile for a single-start right-hand square thread with major diameter 100 mm and lead 36 mm, scale 1:1.

9. Construct the profile for a three-start left-hand Tr 140 × 24 screw thread, scale 1:1 (the designation of this thread is Tr 140 × 72 (P24) LH).

10. Select a suitable thread for the following applications:

(a) general-purpose fastener with risk of vibration
(b) transmitting power load in one direction only
(c) transmitting power load in either direction
(d) general-purpose use
(e) magnesium housing (metric fasteners unavailable)
(f) limited length of engagement.

SECTION D

BEARINGS

Bearing Arrangements and their Use in Industry

After reading this chapter you should be able to:

- ★ know types of bearing arrangement and explain their use in industry (G);
- ★ explain the need for bearings (S);
- ★ identify the most commonly used bearings, for example, ball, roller, plain (S);
- ★ state and explain reasons for the choice of bearing for specified industrial uses (S).

(G) = general TEC objective
(S) = specific TEC objective

INTRODUCTION

In almost all machine constructions, various components either slide or rotate at different speeds while performing a particular function. In order to ensure their security and efficient free running, friction and wear must be reduced to a minimum and this is achieved by the use of bearings.

Generally bearings fall into two classes (Fig. 4.1):

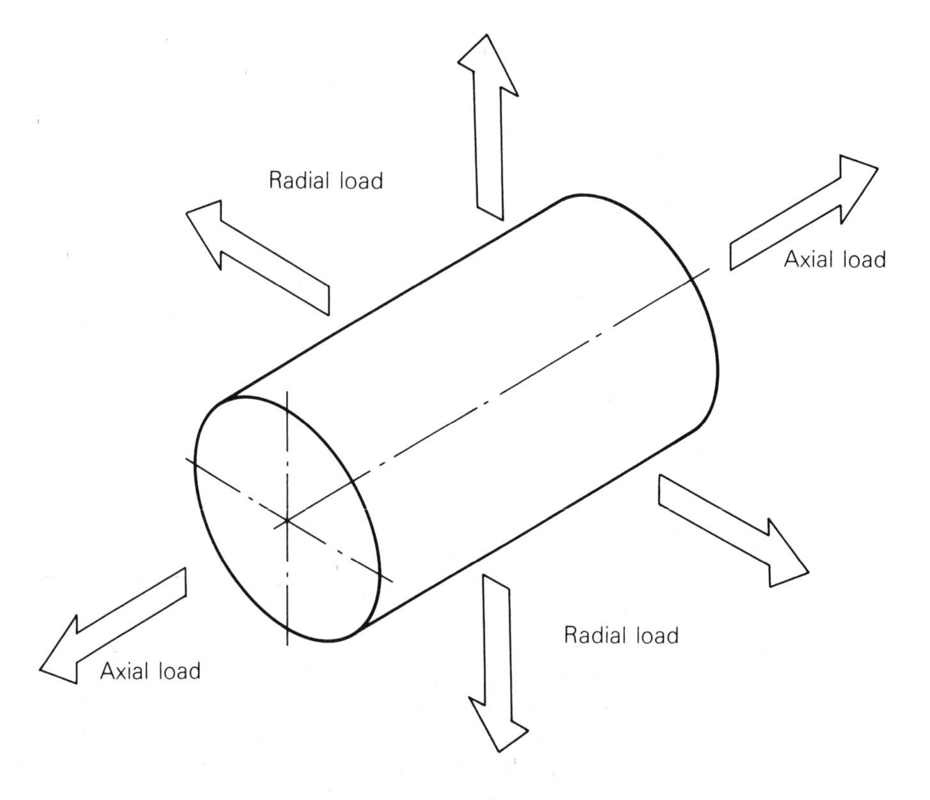

Fig. 4.1 Radial and axial loads on rotating shafts.

(a) radial bearings, which support rotating shafts and mainly radial loads
(b) thrust bearings, which support axial loads on rotating shafts.

TYPES OF BEARING

(a) Plain Bearings

Plain bearings consist of two surfaces which move relative to each other and are usually classified as *sliding bearings*, in which the relative motion is linear (such as a machine tool slideway), or *journal bearings*, in which a shaft rotates inside a journal in the form of a cylinder.

The simplest way of supporting a rotating shaft is by insertion into a block of metal, bored to a suitable size, as shown in Fig. 4.2. This system, however, is unsatisfactory for most applications because there is no allowance for renewal due to wear. It may also only be suitable for supporting shafts running at low speeds due to the friction generated between mating surfaces.

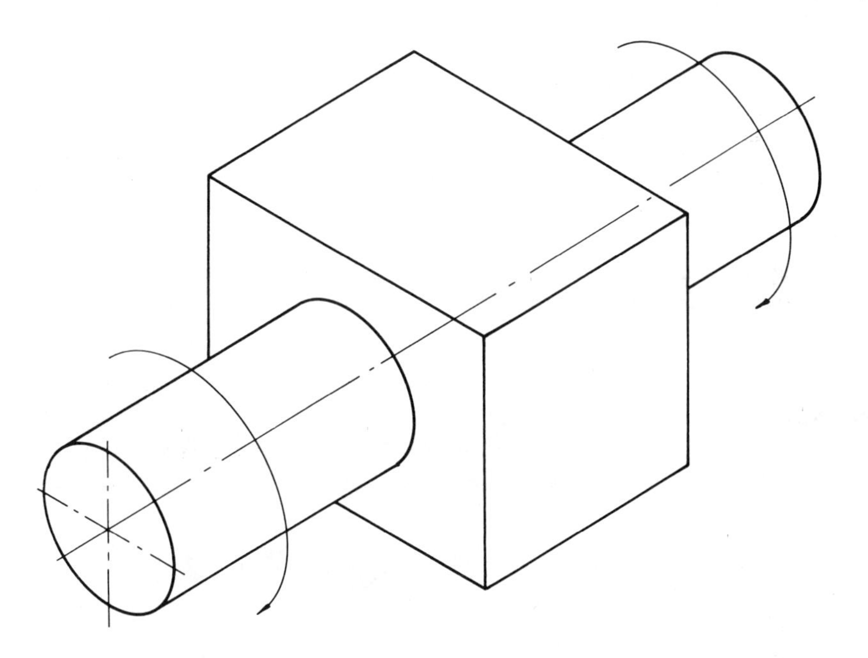

Fig. 4.2 Simple bearing.

Friction could be reduced by using suitable bearing materials, often in the form of a plain or flanged bush (Fig. 4.3), such as phosphor bronze, cast iron, white metal, graphite or plastics.

This type of bearing can be produced at low cost, requires only a small amount of radial space, can accept small shock or overload conditions and is quiet in operation. The outside diameter is a press fit into the housing and the bore is slightly larger than the shaft to allow for rotation. After it has become worn beyond further use it can be removed and replaced quite easily.

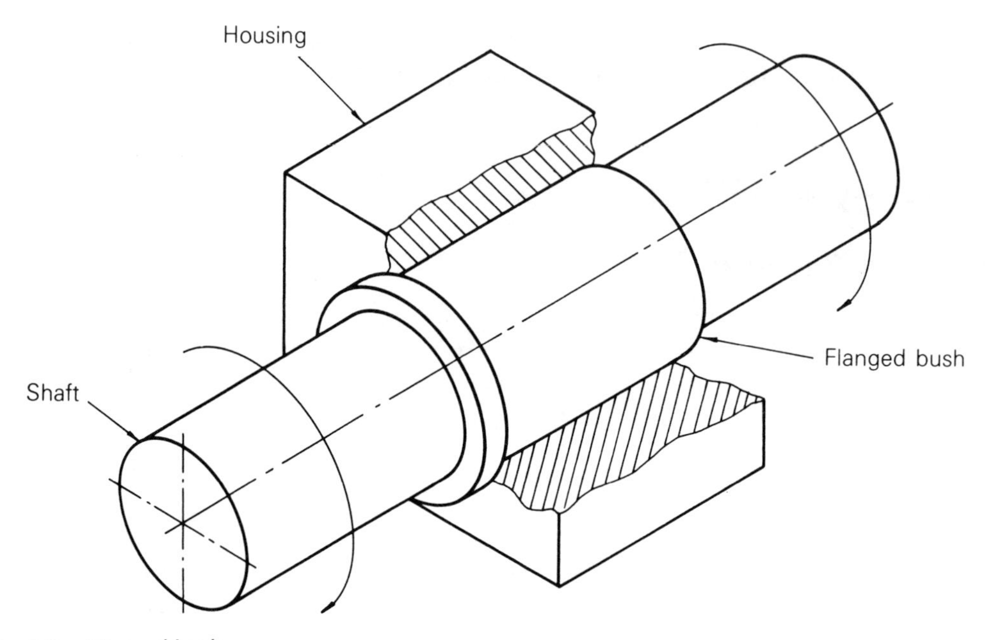

Fig. 4.3 Flanged bush.

(b) Rolling Bearings

The use of plain bearings has been in evidence since the invention of the wheel, but in the latter part of the nineteenth century the introduction of the bicycle and the advent of the internal combustion engine led to the development of the ball and roller bearing industry.

The rolling components in bearings take many shapes—spheres, cylinders, spherical cylinders and cones—but the shape which can be most easily and accurately produced is the sphere. Consequently the ball bearing is the most widely used type of bearing.

Ball bearings

A radial ball bearing is designed to take radial loads but is also capable of accepting small axial loads or a combination of both—see Fig. 4.4.

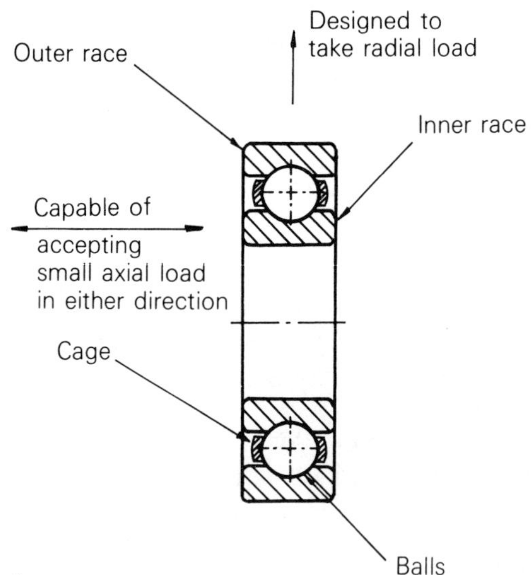

Fig. 4.4 Radial ball bearing.

Thrust bearings (Fig. 4.5) are not designed to carry any radial load. They are particularly used for conditions of heavy axial load at low speed.

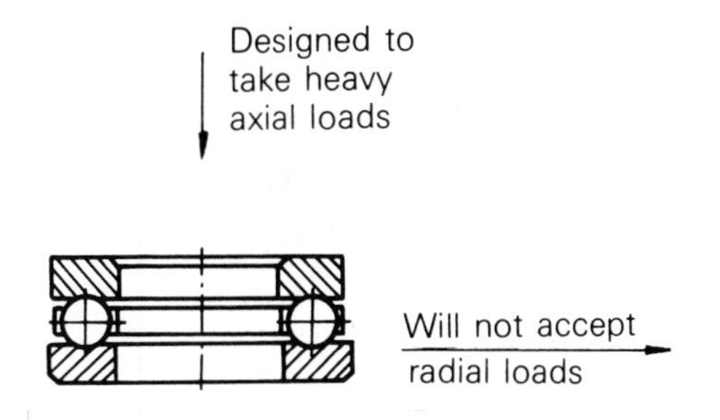

Fig. 4.5 Thrust ball bearing.

Other types of ball bearing include the single-row angular-contact bearing shown in Fig. 4.6. The radial and axial load capacities are higher than those of the radial ball bearing although it can be seen from the construction that an angular-contact bearing can support load in one direction only. Angular-contact bearings are often used in pairs adjusted one against another, when they will carry axial load in either direction, radial load or a combination of the two. Double-row angular-contact bearings are also available which can achieve the load capacity of two single-row bearings in face-to-face or back-to-back configurations.

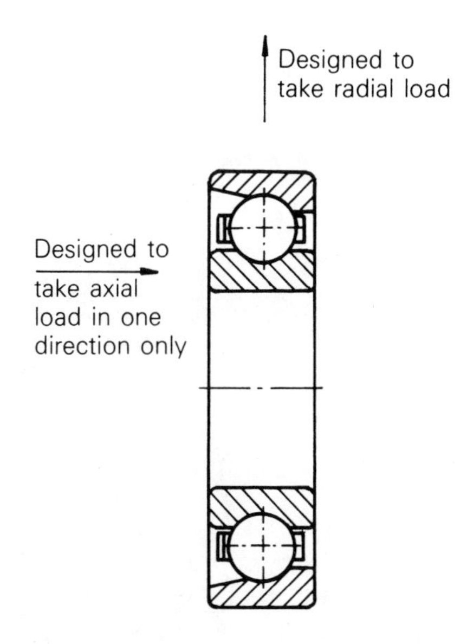

Fig. 4.6 Single-row angular-contact bearing.

Double-row self-aligning ball bearings are used where accurate alignment cannot be guaranteed. They will carry radial loads but axial loads should be avoided (Fig. 4.7). They are not suitable for shock or vibration.

Roller bearings

Roller bearings have greater radial load-carrying capacity than ball bearings of the same external dimensions, and are suitable for arduous duty. The general type of roller bearing, shown in Fig. 4.8, is not suitable for axial loading.

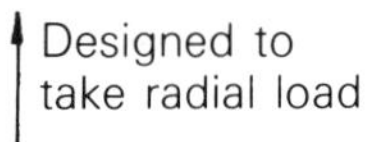

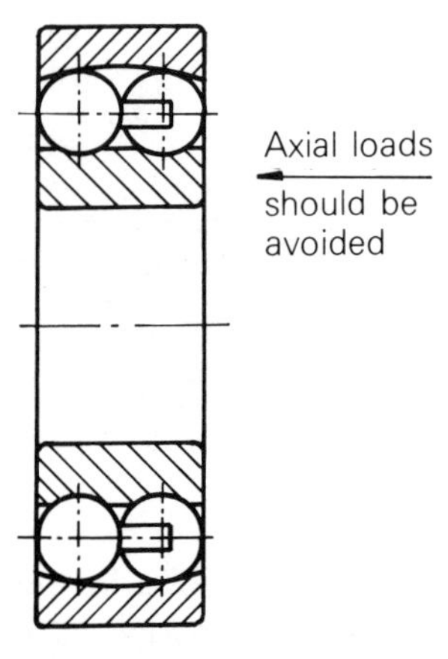

Fig. 4.7 Double-row self-aligning ball bearing.

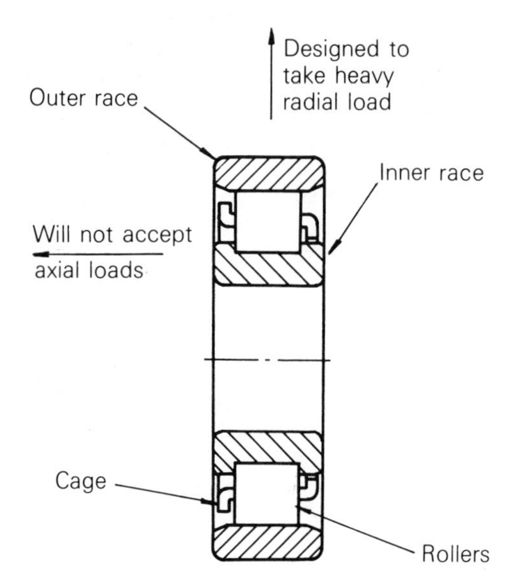

Fig. 4.8 Roller bearing.

Needle roller bearings (Fig. 4.9) are fitted with rollers whose length is much greater than their diameter. These types of bearing will support radial loads and may be supplied without an inner race, as shown in Fig. 4.9a, to run direct on a hardened steel spindle. A cheaper, very compact, type of needle roller bearing is shown in Fig. 4.9b and incorporates a number of needle rollers retained by a plastics or steel cage.

Fig. 4.10 shows a double-row spherical roller bearing. These bearings are similar in principle to the double-row self-aligning ball bearings, but are capable of carrying appreciably higher radial loading. They also carry limited axial loading.

Tapered roller bearings (Fig. 4.11) are designed to carry both radial and axial loads. Like angular-contact bearings they are often used in pairs adjusted one against another, when they will carry axial loads in either direction. They are made with various angles and roller lengths in order to suit the radial or axial load requirements of a given application. Axial loading can be increased with a steeper taper angle.

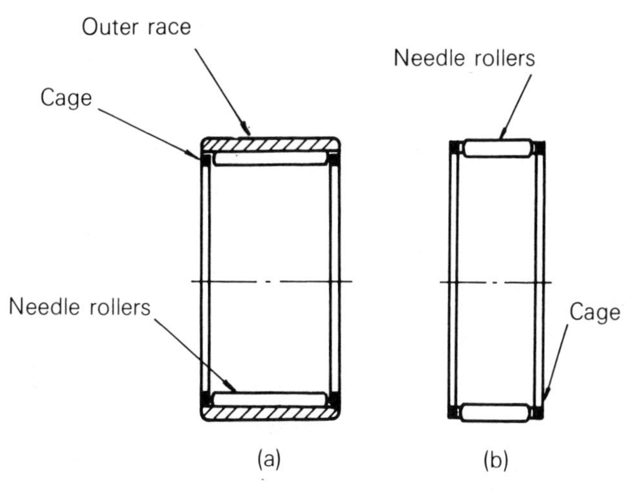

Fig. 4.9 Needle roller bearings.

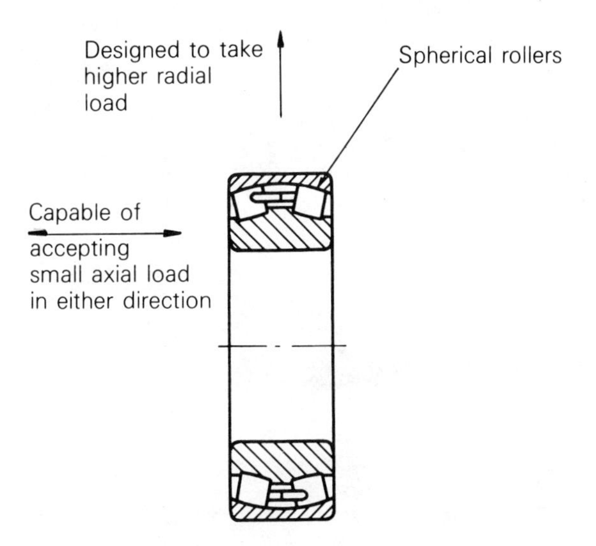

Fig. 4.10 Double-row spherical roller bearing.

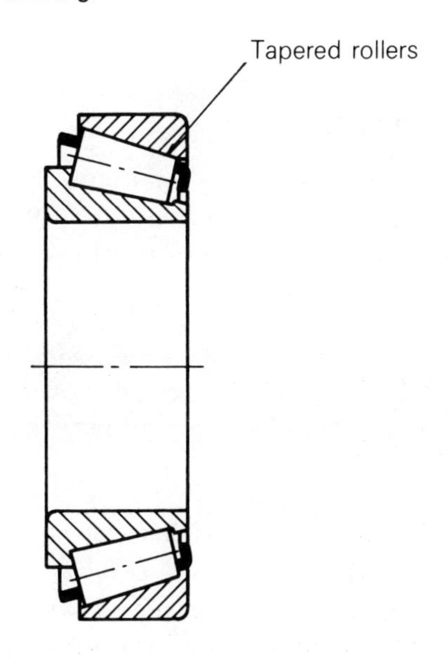

Fig. 4.11 Tapered roller bearing.

Fig. 4.12 shows a range of tapered roller bearings from small single-row general-purpose types to multi-row bearings. Fig. 4.13 shows a four-row bearing used mainly in the rolling mill industry.

Fig. 4.12 A range of tapered roller bearings. (By courtesy of The Timken Co.)

CHOICE OF BEARING

(a) Plain Journal Bearings

These bearings are most commonly used in industry because they have a low initial cost and are particularly suitable where shock or overloading conditions may occur. They need to have the following characteristics.

Wear resistance

Once installed, the bearing needs to be in service for a long time to maintain efficiency. Although a plain bearing is produced at low cost, the resulting loss of production from any breakdown can be very expensive.

Low coefficient of friction

Energy will be wasted in overcoming friction and will result in the generation of heat. Eventually the bearing may overheat and seize up. Friction may be reduced by:

(i) lubrication, which separates the spindle and bearing by a thin film of oil and also helps to cool the assembly

Fig. 4.13 A four-row tapered roller bearing. (By courtesy of The Timken Co.)

(ii) producing mating parts which have a smooth surface texture and thereby limiting the cause of friction forces

(iii) using materials with low coefficient of friction values.

Plain journal bearings are compact in design, needing only a limited amount of radial space and are used for a wide range of applications. The following materials are used for these bearings.

Phosphor bronze

This has good anti-friction properties and is suitable for heavy loads because of its rigidity. It is used for transmission bearings, big-end bearings for connecting rods and general-purpose applications. Bronze may be alloyed with lead (leaded bronze) to the exclusion of phosphorus to improve the lubrication properties of the bearing.

Cast iron

The flakes of graphite in grey cast iron provide excellent lubrication and are an added advantage for the use of this metal as a bearing material. It is used for medium loads at moderate speeds, for example, machine tool slideways and spindle journals.

White metal

White metal alloys contain copper, tin, antimony or lead. Originally white metal was cast around an in-situ mandrel (which was slightly smaller than the shaft to be supported) and after cooling it was scraped to size. Modern practice is to use the white metal as a thin coating on a stronger metal shell so that renewal can be effected by replacing a complete shell.

Babbit metal, a tin-based white metal, has a low melting point and must be well lubricated. It is often used for big-end and main bearings in automobile engines.

Graphite

Graphite (carbon) bearings are self-lubricating and are particularly suitable for high-temperature applications, such as furnace and boiler equipment.

Porous bearings

These are made from powdered metal (often bronze) which is pressed to shape and then sintered. They are able to absorb oil within the structure and are used where lubrication supply is difficult or not expected, for example, machine tool and motor vehicle parts.

Plastics bearings

These are made from nylon and PTFE and can be produced at low cost in large quantities by moulding. They have low coefficients of friction and are resistant to corrosion which, together with their cleanliness, makes them ideal for domestic appliances, such as food mixers and washing machines, and office machinery.

(b) Rolling Bearings

The rolling elements (balls or rollers) and races are hardened and ground to close limits and consequently frictional resistance is reduced. Because friction is reduced, less power is required. Under normal operating conditions (which does not include shock loading) wear is almost negligible, and although lubrication is not required for the bearing property it is used to prevent corrosion. A standard range of bearings is available with masses varying from a few grams to several tonnes.

When selecting a bearing it must be determined just what is required of it in use:

What are the maximum radial and axial forces the bearing must support?

What speeds will the bearing be operating at?

Can accurate alignment be maintained between bearing housings?

What is the nature of the environment in which the bearing will be operating?

Will the presence of dirt or moisture require special consideration?

Will the bearing be subject to high-temperature working?

It is important to select the correct bearing for the application. In some cases bearing loading may be indeterminate and the best results may be obtained by basing selection on the known results of similar applications. Bearing manufacturers, however, have a wealth of information and experience of bearings used in widely differing conditions and it is strongly recommended that an approach should be made to them, for they will be only too willing to offer their advice on selection for a given application.

The mounting of bearings is very important. Fig. 4.14 shows a typical mounting for two single-row radial ball bearings used to locate a rotating shaft axially in both directions at one position. Some end movement is inevitable due to the internal construction of the bearing, but this can usually be accommodated. The bearing inner races are interference fits, (for example, press fits) and the outer races are clearance fits (for example, sliding fits) in the housings. One bearing is clamped

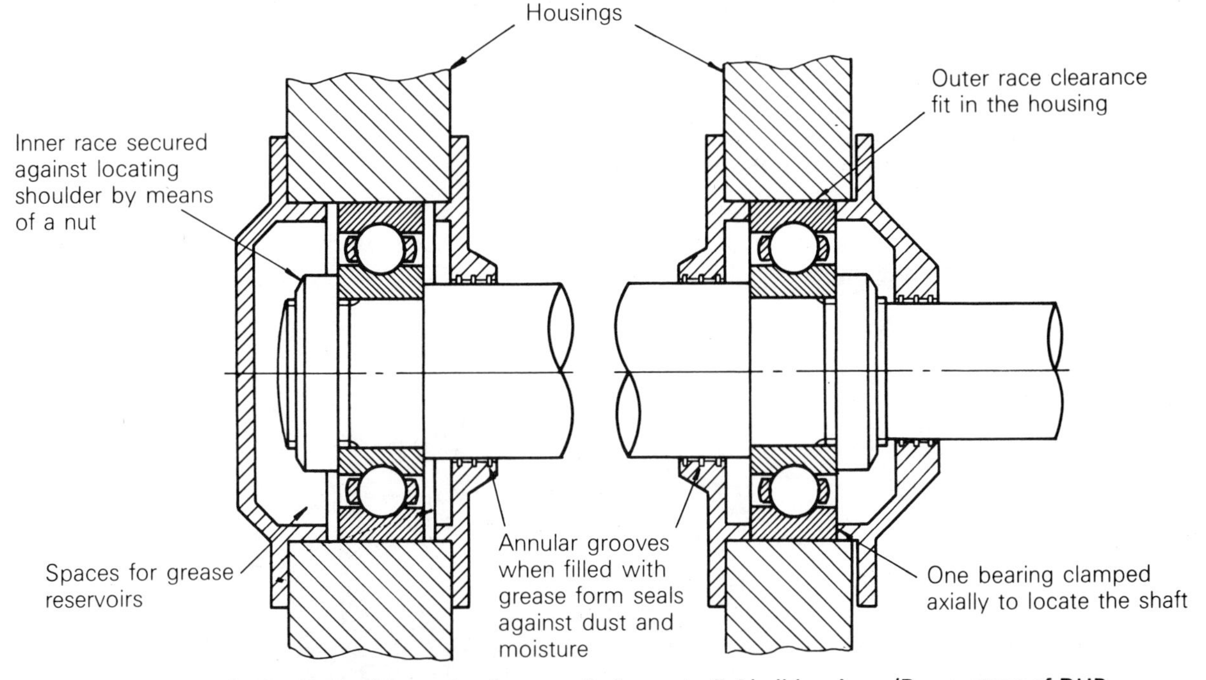

Fig. 4.14 Typical mounting for two single-row radial ball bearings. (By courtesy of RHP Bearings Ltd.)

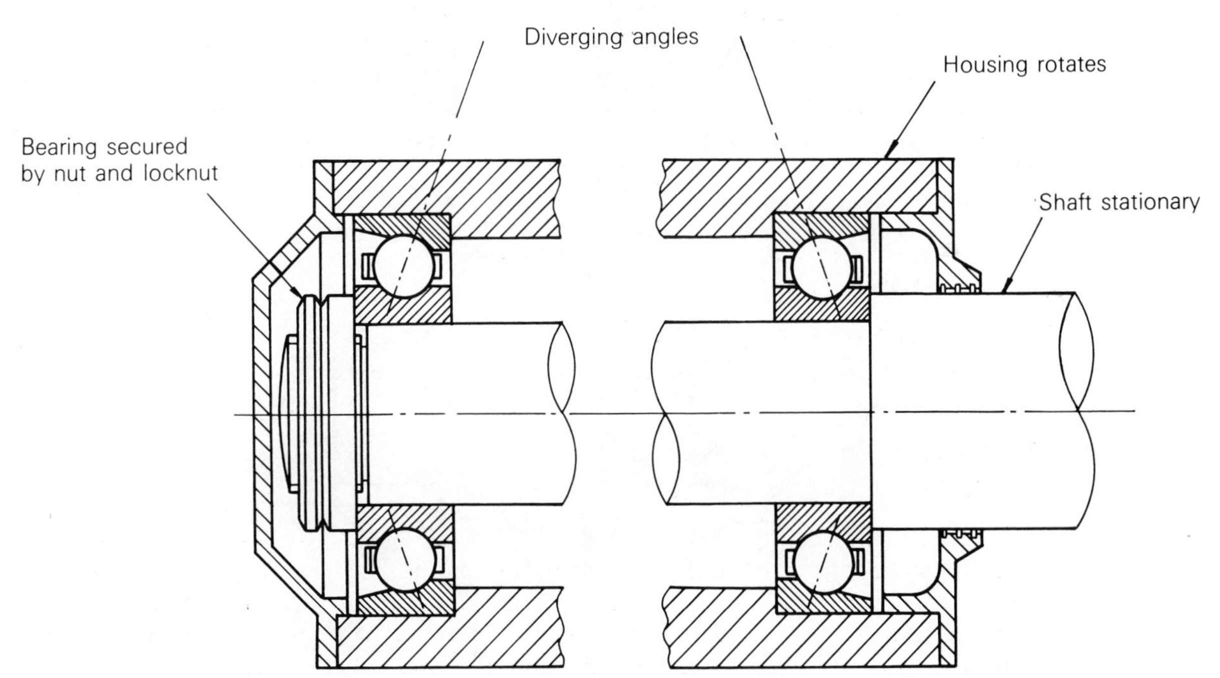

Fig. 4.15 Typical back-to-back mounting for two single-row angular contact bearings. (By courtesy of RHP Bearings Ltd.)

axially and locates the shaft, the outer race of the other bearing is free to avoid any possibility of the bearings being forced together axially. The inner race is held up against a locating shoulder by means of a nut. Spaces on each side of the bearings form a grease reservoir and the annular grooves in the bores of the end covers, when filled with grease, form effective sealing against the ingress of dust and moisture.

Fig. 4.15 shows a typical back-to-back mounting for two single-row angular contact bearings. In this arrangement the diverging contact angles of the bearings give a rigid assembly. Adjustment is carried out through one of the inner races, which must be a sliding fit on the shaft. Consequently this arrangement should only be used when the bearing housing rotates and the shaft is stationary. Allowance for any thermal expansion must be made in the adjustment. The bearings can be axially preloaded but any overload condition should be avoided.

Fig. 4.16 shows a single-direction thrust ball bearing used in conjunction with cylindrical roller bearings which is suitable for heavy downward axial loading where accurate alignment can be guaranteed between the two bearing housings. The bore

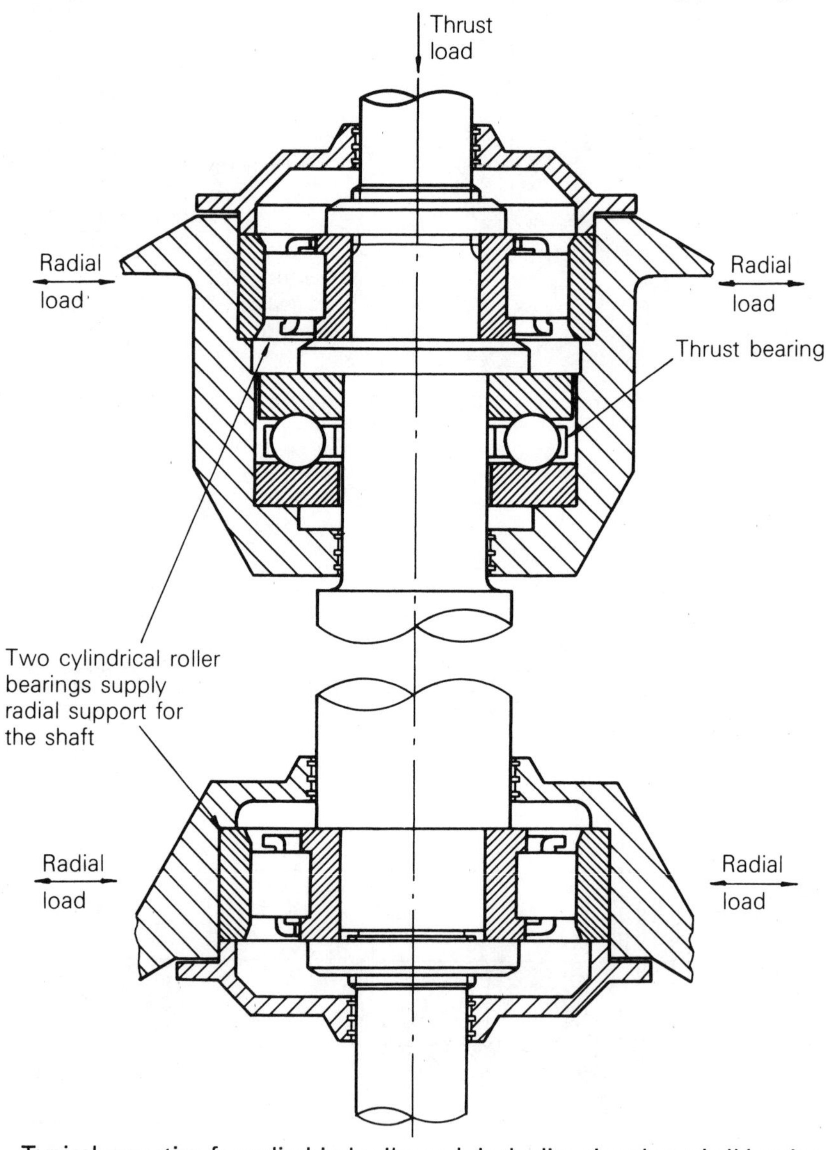

Fig. 4.16 Typical mounting for cylindrical roller and single-direction thrust ball bearings. (By courtesy of RHP Bearings Ltd.)

of the small washer and the outside diameter of the large washer of the thrust bearing are manufactured to close limits for fitting to the shaft and housing seatings respectively. Care is required to avoid fitting these bearings the wrong way round because they would not then be properly centralised. The roller bearings supply the radial support for the shaft.

Table 4.1 *Applications of rolling bearings*

Type of Bearing	*Design Advantages*	*Design Limitations*	*Typical Use*
Radial ball	Relatively cheap to produce. Accept radial loads	Will only accept small axial loads	Cycles, electric motors, general use
Thrust ball	Accept heavy axial load	Will not accept radial load. Superseded on high-speed applications	Crane hooks and pivots
Single-row angular-contact ball	Accept radial load and axial load in one direction only. Will accept axial load in both directions when paired	If used singly, axial load must exceed radial load at all times	Spindles in machine tool applications
Double-row self-aligning ball	Accommodate initial misalignment	Limited load-carrying capacity	Where accurate alignment cannot be guaranteed
Roller	Higher capacity for radial loading than equivalent size radial ball bearing	In general form not suitable for axial loading	Heavily loaded spindles in machine tools and gear boxes
Needle roller	Compact and light weight	Not suitable for axial loading	Where excess weight is undesirable, e.g. aircraft
Double-row spherical roller	Similar in principle to double-row self-aligning ball bearings but capable of carrying higher radial loading	Speed capabilities are not high	Where accurate alignment cannot be guaranteed
Tapered roller	Accept both radial and axial loading	If used singly, axial load must be in one direction only	Motor car front hubs, railway axle boxes

SELF-TEST 4

Select the most suitable option(s). *Note*: There may be more than one correct answer.

1. The plain bearing most suitable for additional small axial load is the:

(a) plain bush
(b) flanged bush
(c) journal bearing
(d) journal.

2. The rolling bearing which can be most easily and accurately produced is the:

(a) ball bearing
(b) roller bearing
(c) spherical roller bearing
(d) taper roller bearing.

3. A radial ball bearing is designed to take:

(a) light radial and light axial loads
(b) medium radial and medium axial loads
(c) light radial and medium axial loads
(d) medium radial and light axial loads.

4. The most suitable bearing for shock loading is the:

(a) ball bearing
(b) roller bearing
(c) thrust bearing
(d) plain bearing.

5. The most suitable bearing to carry both radial and axial loads is the:

(a) tapered roller bearing
(b) needle roller bearing
(c) plain roller bearing
(d) spherical roller bearing.

6. The factors determining the suitability of a material for plain bearings are:

I low friction properties; II resistance to wear; III ability to withstand emergency conditions without seizure:

(a) I and II only
(b) II and III only
(c) I, II and III
(d) I and III only.

7. Bronze porous bushes are generally used for light load conditions. The load can be increased to substantial values if:

I speed is decreased; II load is intermittent:

(a) neither of these
(b) I only
(c) II only
(d) I and II.

8. Which of the following bearing materials is (are) self-lubricating?

(a) cast iron
(b) white metal
(c) graphite
(d) phosphor bronze.

9. The most suitable combination to sustain heavy radial and light axial loading is:

(a) a pair of angular contact bearings
(b) a radial ball bearing and a roller bearing
(c) a radial ball bearing and a thrust bearing
(d) a pair of double-row spherical roller bearings.

10. The most suitable material from which to make a plain bearing to support heavy radial loads is:

(a) cast iron
(b) phosphor bronze
(c) white metal
(d) powdered bronze.

EXERCISE 4

1. State and describe the two main classes of bearing.

2. Sketch and describe the most widely used rolling bearing.

3. How is wear resistance in nylon bushes enhanced?

4. Which type or combination of rolling bearing(s) would be most suitable for the following applications?

(a) heavy axial load in one direction only
(b) heavy radial and axial load at moderate speed
(c) medium axial loading in two directions
(d) heavy radial loading only
(e) medium radial loading together with small axial loading.

5. State three characteristics required by a plain bearing.

6. What advantage does a surrounding medium, for example, water, alkalis, acids, have on plastics bearings?

7. Nylon and PTFE are two plastics materials used for bearings. Name two other types of plastics which are used for the same purpose.

8. Give one reason why porous bearings are used for applications where the bearings would be inaccessible.

9. Describe, with the aid of sketches, the difference between a plain bush and a flanged bush. State three metallic materials from which bushes are made and give one advantage and one use of each.

10. Fig. 4.14 shows a typical mounting for two single-row radial ball bearings. Re-draw the arrangement, replacing the bearings with two bearings which will additionally support a moderate axial load in either direction.

FURTHER READING

T.S. Nisbet, *Rolling Bearings* (Oxford University Press, 1974).

SECTION E

GEOMETRICAL TOLERANCING

The Tolerances of Geometry and Size

After reading this chapter you should be able to:

- ★ understand that components, in addition to errors of size, can have errors due to form, attitude and location and relate geometrical tolerancing to BS 308: Part 3: 1972 (G);
- ★ define a geometrical tolerance (S);
- ★ state the reasons for using geometrical tolerances (S);
- ★ state the general principles of geometrical tolerances (S);
- ★ identify the recommended symbols for tolerancing form — straightness, flatness, roundness, cylindricity, profile of a line, and profile of a surface (S);
- ★ identify the ISO recommended symbols for tolerancing attitude — parallelism, squareness and angularity (S);
- ★ identify the recommended symbols for tolerancing location — position, concentricity and symmetry (S);
- ★ identify clearance, interferance and transition fits used in BS 4500A and explain their application to practical problems (S);
- ★ dimension drawings in accordance with BS 308: Part 2: 1972 (S).

(G) = general TEC objective
(S) = specific TEC objective

INTRODUCTION

A geometrical tolerance controls the maximum permissible overall variation of form or position of a feature, i.e. it defines the size and shape of a tolerance zone within which the surface or axis of the feature is to lie. It represents the full indicator movement in cases where testing with an indicator is applicable.

Fig. 5.1 shows examples of geometrical tolerances. Fig. 5.1a shows a spindle for which a straightness of 0.05 mm is required. This means that the axis of that part of the spindle to be controlled is required to be contained within a cylindrical zone of diameter 0.05 mm. In Fig. 5.1b the surface indicated is required to lie between two parallel planes 0.05 mm apart. Fig. 5.1c shows a roundness tolerance. The periphery at any cross-section perpendicular to the axis is required to lie between two circles concentric with each other, a radial distance 0.05 mm apart, in the plane of the section.

WHY USE A GEOMETRICAL TOLERANCE?

The student will be aware from previous studies that a component cannot be manufactured to an exact size. If a spindle is to be machined to a diameter of 25 mm it will, most likely, be slightly oversize or slightly undersize — it is impossible to achieve an exact size even with the use of the most advanced machining facilities.

A designer must be able to accommodate some enforced error in every feature of each component which is to be manufactured. The amount of error to be tolerated will depend on the function of each feature. Every dimension on an engineering drawing should be toleranced in order that components can be made and assembled to perform their duty adequately. Often a drawing will specify an overall tolerance which is applicable to all dimensions unless otherwise stated.

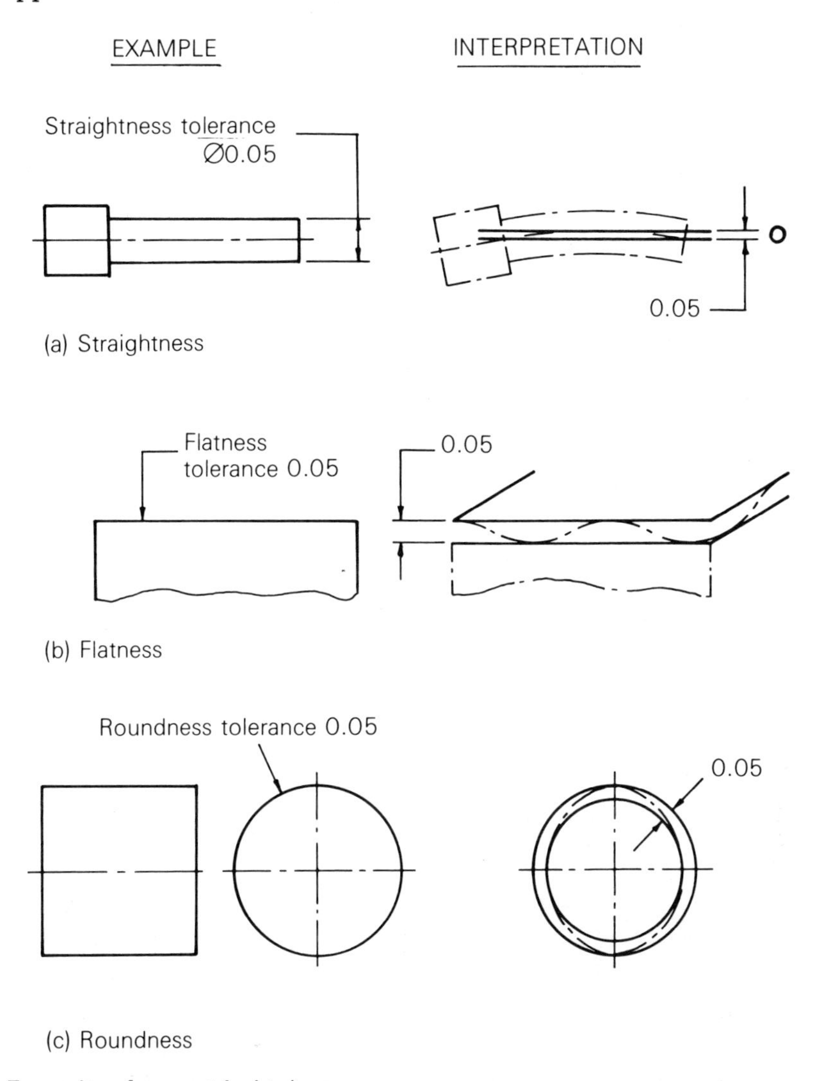

Fig. 5.1 Examples of geometrical tolerances.

In addition to errors of size, a component may be subject to an error of form, attitude or location. Consider the spindle shown in Fig. 5.2. In Fig. 5.2a the spindle is drawn and the feature being considered — the largest diameter — is dimensioned and toleranced such that any size between 11.03 mm diameter and 11.05 mm diameter is acceptable. In Fig. 5.2b it can readily be seen that the feature is too large — an error of size.

In Figs 5.2c, d and e the feature is within the size tolerance specified but the component is unacceptable due to geometrical errors. In Fig. 5.2c the feature has an error of form — it is lobed. (Lobing is a geometrical error which often occurs in a centreless grinding operation. It is not detectable by measurement because the size across its extremities is always the same, giving the impression of roundness.) In Fig. 5.2d there is an error of attitude — the feature is not square with the end face of the smaller diameter. In Fig. 5.2e the feature has an error of location — it is not concentric with the smaller diameter.

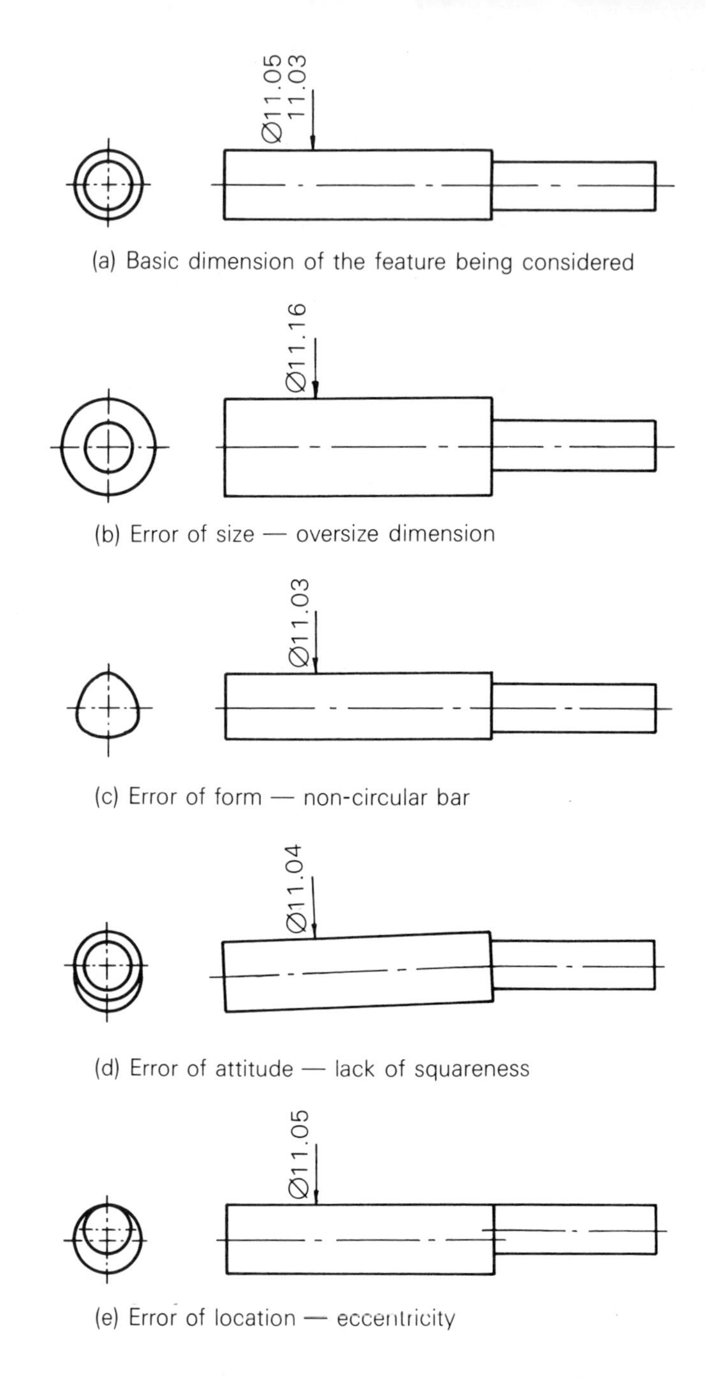

Fig. 5.2 Types of error.

How far it is necessary to specify geometrical tolerances in any particular instance can only be decided in the light of functional requirements and probable manufacturing circumstances. Drawings prepared for widespread quantity production may need the most complete and explicit dimensional and geometrical tolerancing so that the part may be made, and inspected, to suit the full requirements of the designer. On the other hand, such detail may be unnecessary when adequate control is exercised by other means, for example, when the method of production has been proved to produce parts to the required tolerances for satisfactory functioning.

Unless otherwise stated, a geometrical tolerance applies to the whole length or surface of a part. (Fig. 5.3 shows a tolerance restricted to a particular part of a component.)

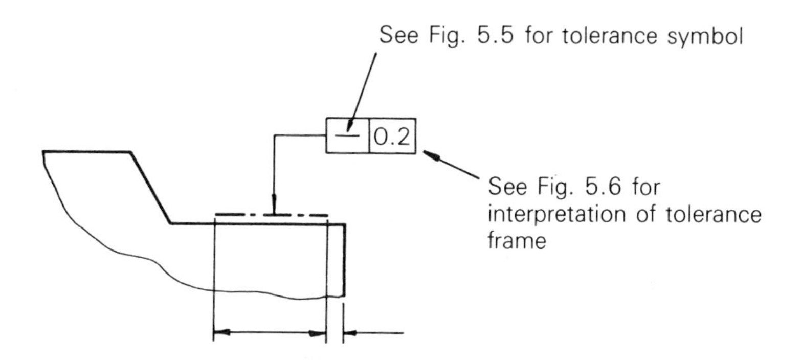

Fig. 5.3 Tolerance applicable to restricted length of feature.

When geometrical tolerances are expressed they are to be observed regardless of the actual finished sizes of the parts concerned.

A surface or feature may be of any form or may take up any position within the tolerance zone specified, except where a further restriction is imposed. An example of this is shown in Fig. 5.4, where the surface of a component is required to lie between two parallel planes 0.05 mm apart. The interpretation of the tolerance is that the surface may be of any shape but must not be concave.

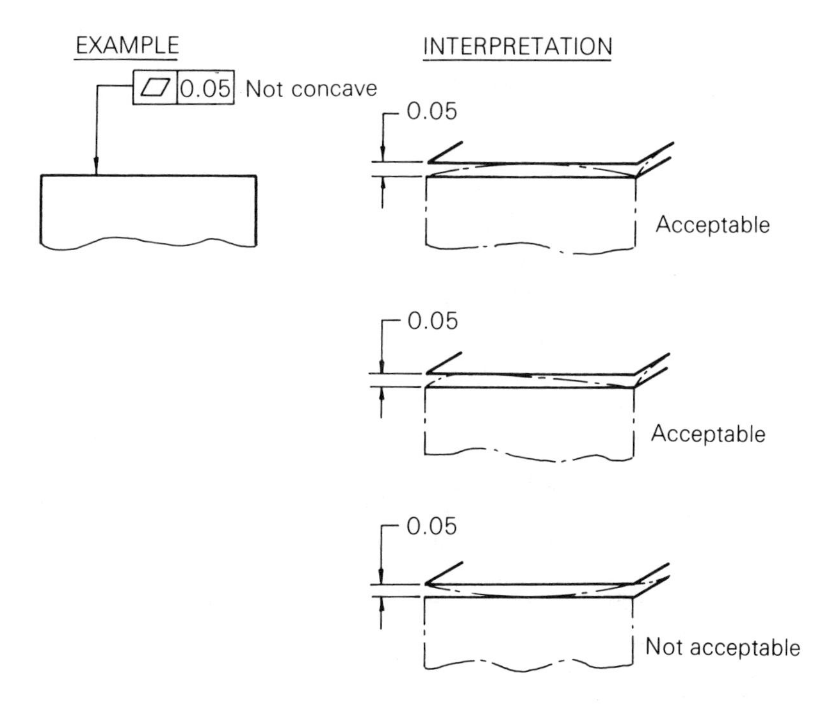

Fig. 5.4 Tolerance restriction.

The student should appreciate, while studying the following sections, that in spite of the various terms (straightness, flatness, roundness, etc.) used to describe geometrical tolerances, it often happens that one type of geometrical tolerance will automatically limit other types of geometrical error. For example, the parallelism quoted in Fig. 5.16b, in requiring the surface to be contained within a tolerance zone bounded by two planes 0.2 mm apart, limits errors of flatness as well as those of parallelism.

Certain types of geometrical tolerance require a feature other than that being toleranced to be indicated as a datum. The use of a feature as a datum, an example of which is shown in Fig. 5.17, requires that it should itself have an adequate accuracy of form or position and it may be necessary in some cases to specify tolerances of form for datum features. Datum features should be chosen from considerations of the function of the part. In some cases, however, it may also be desirable to indicate the position of certain points intended to form a temporary datum feature for both manufacture and inspection. (The use of a feature as a datum and the principle of specifying a datum are shown in Figs. 5.6, 5.7 and 5.8.)

SYMBOLS FOR TOLERANCING FORM

The symbols shown in Fig. 5.5 are used to indicate the characteristics to be toleranced.

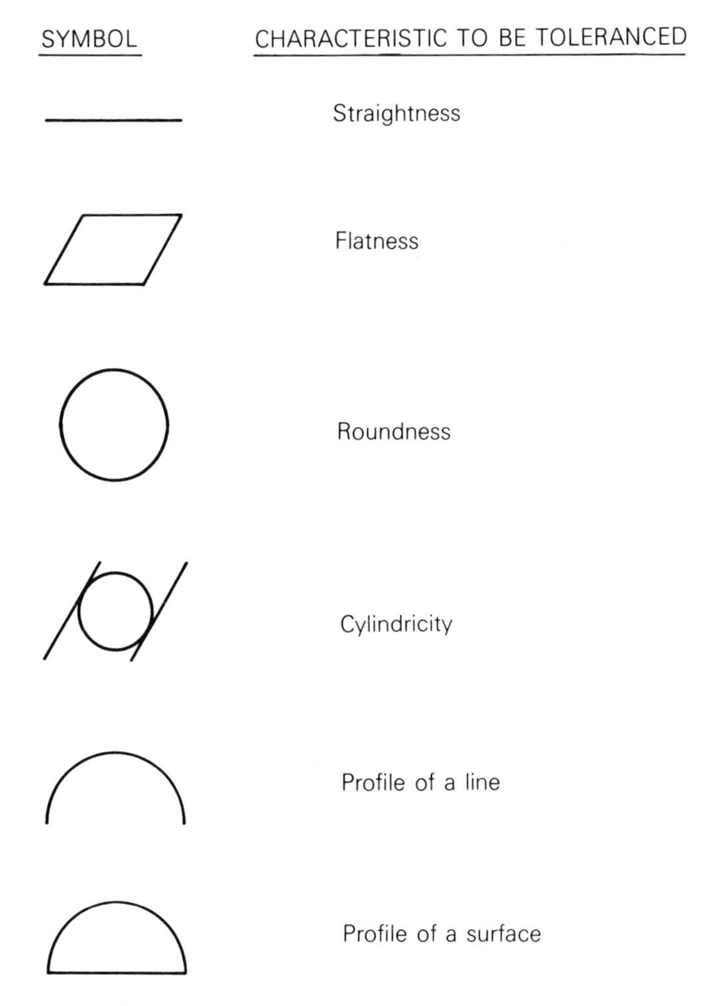

Fig. 5.5 Symbols for tolerancing form.

The geometrical tolerance is indicated in a rectangular frame which is divided into compartments as shown in Fig. 5.6.

The symbol for the characteristic (see Fig. 5.5) being toleranced is shown in the left-hand compartment (Fig. 5.6a).

The total value of the tolerance, generally in millimetres, is shown in the second compartment from the left (Fig. 5.6b). This value is preceded by the sign ∅ if the total tolerance zone is circular or cylindrical.

In cases where a datum system is used, the third compartment is used for the datum identification letters (Fig. 5.6c).

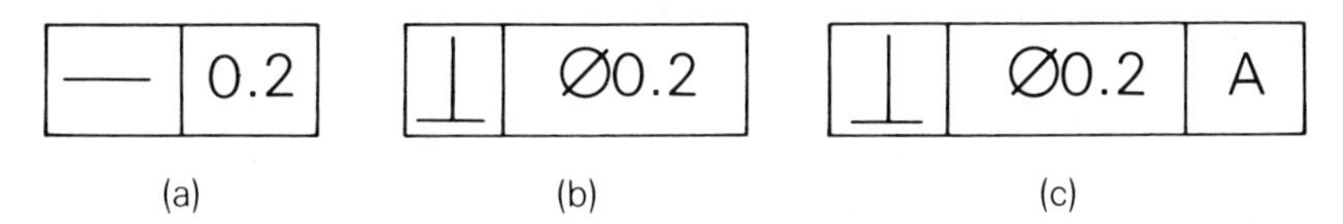

Fig. 5.6 Tolerance frame.

The feature controlled by the tolerance is indicated by a leader line connecting it to the tolerance frame, as shown in Fig. 5.7. At the toleranced feature the leader line terminates in an arrowhead which may be positioned:

(a) on the outline of a feature when the tolerance refers to the surface represented by a line (Fig. 5.7a).
(b) on a projection line of a feature at a dimension line when the tolerance refers only to the axis of the feature being dimensioned (Fig. 5.7b).
(c) on the axis when the tolerance refers to the common axis of all features lying on it (Fig. 5.7c).

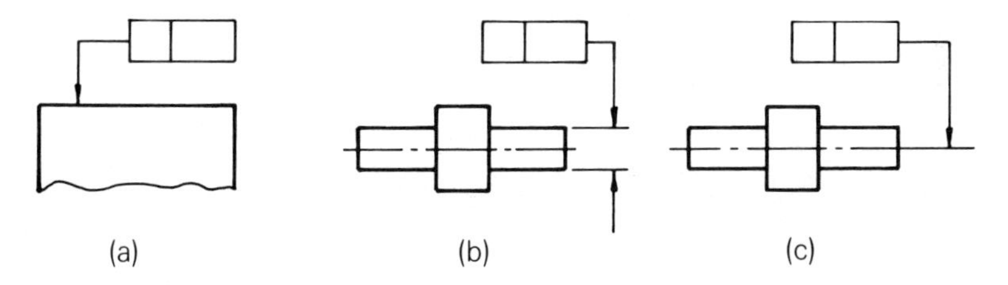

Fig. 5.7 The feature controlled.

A datum feature, as shown in Fig. 5.8, is indicated by a leader line from the tolerance frame terminating in a solid triangle the base of which lies:

(a) on the outline of the feature when the datum feature is the surface itself (Fig. 5.8a)
(b) on a projection line at the dimension line when the datum feature is the axis of the part so dimensioned (Fig. 5.8b)
(c) on the common axis of two or more features (Fig. 5.8c).

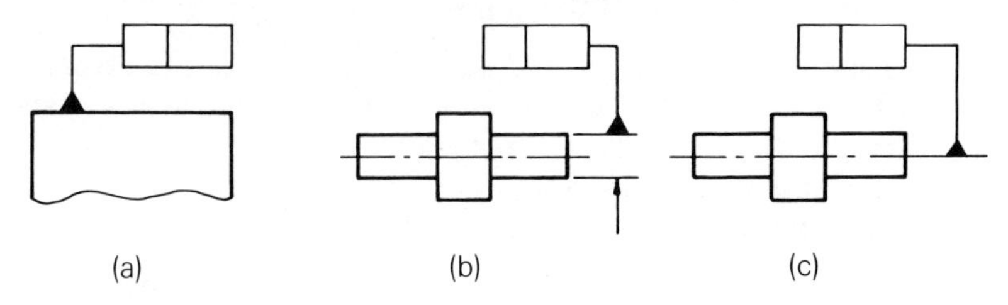

Fig. 5.8 Datum features.

Examples and interpretations of tolerances of straightness are shown in Fig. 5.9.

In Fig. 5.9a, a scribed line shown on the surface is required to lie between two parallel straight lines on the surface, 0.02 mm apart.

In Fig. 5.9b, the surface of the feature in any of its positions is required to lie between two parallel straight lines, 0.04 mm apart, lying in an axial plane.

In Fig. 5.9c, the axis of that part of the piece to be controlled is required to be contained in a cylindrical zone of diameter 0.04 mm.

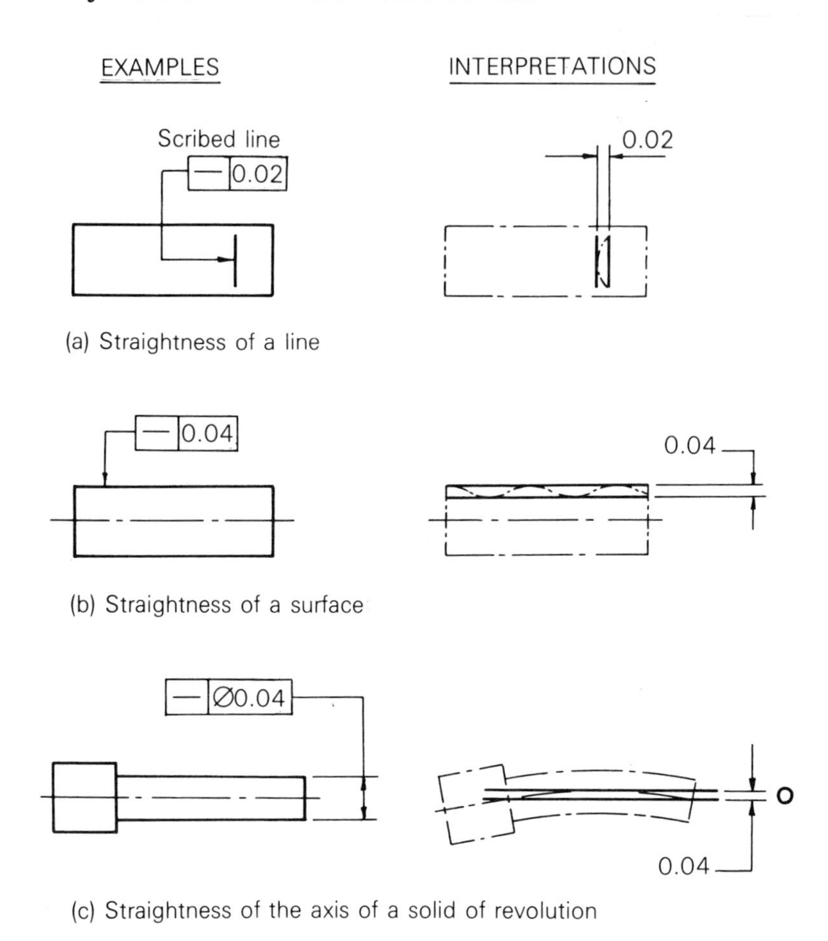

Fig. 5.9 Tolerances of straightness.

A tolerance of flatness is used to control the flatness of a surface. The tolerance zone is the space between two parallel planes and the tolerance value is the distance between the two planes. The indicated surface in Fig. 5.10 is required to lie between two parallel planes 0.05 mm apart.

A roundness tolerance may be used to control the errors of form of a circle in the plane in which it lies. For a cylinder (Fig. 5.11) the plane is perpendicular to the axis. In this example, the periphery at any cross-section is required to lie between two circles concentric with each other, a radial distance 0.01 mm apart, in the plane of the section.

Note: a roundness tolerance is not concerned with the position of the circle, for example its concentricity with the datum axis.

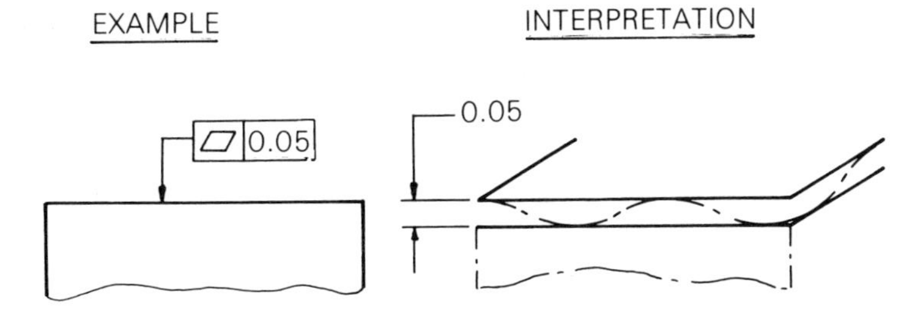

Fig. 5.10 Tolerance of flatness.

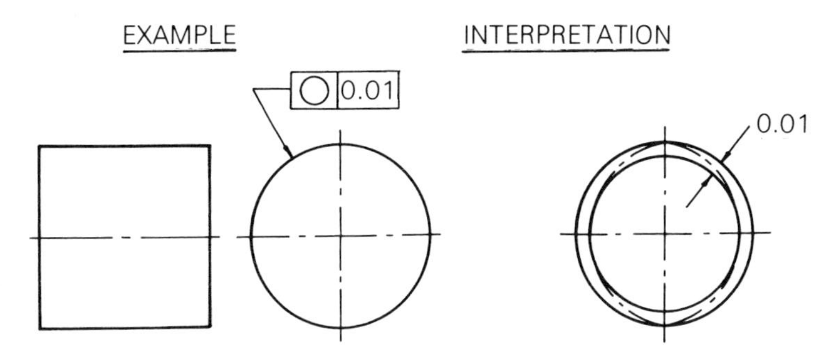

Fig. 5.11 Tolerance of roundness.

Cylindricity is a combination of roundness, straightness and parallelism applied to the surface of a cylinder.

Note: The end surfaces of a cylindrical part are not controlled by a cylindricity tolerance. Fig. 5.12 shows an example of this type of tolerance with its application. The curved surface of the part is required to lie between two cylindrical surfaces coaxial with each other, a radial distance 0.03 mm apart.

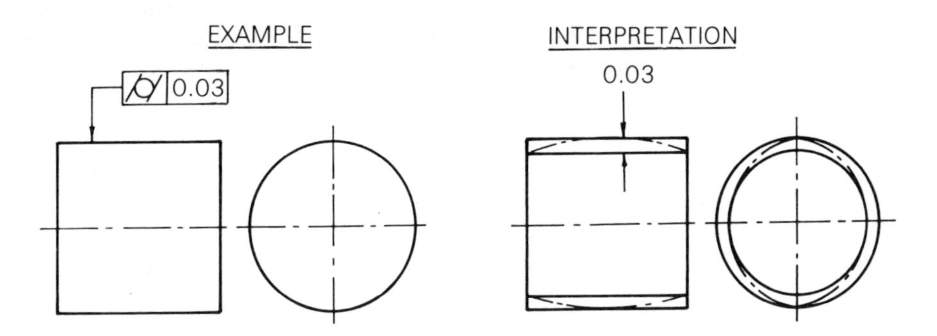

Fig. 5.12 Tolerance of cylindricity.

When the profile of a line is toleranced, the theoretical form of the profile is defined by boxed dimensions and a tolerance zone is established in relation to it. This gives a tolerance zone with a constant width normal to the theoretical profile. A unilateral tolerance zone is shown in Fig. 5.13. (A tolerance is unilateral when the limits are both above or both below the nominal size.) In any section parallel to the plane of projection of the drawing, the actual profile is required to lie between the theoretical profile and a line which envelops a series of circles of diameter 0.2 mm, touching and inside the theoretical profile.

Unless the drawing indicates a unilateral tolerance (as in Fig. 5.13), an equally disposed bilateral tolerance is implied (a tolerance is bilateral when one limit is above and the other below the nominal size) and the thick chain line is omitted.

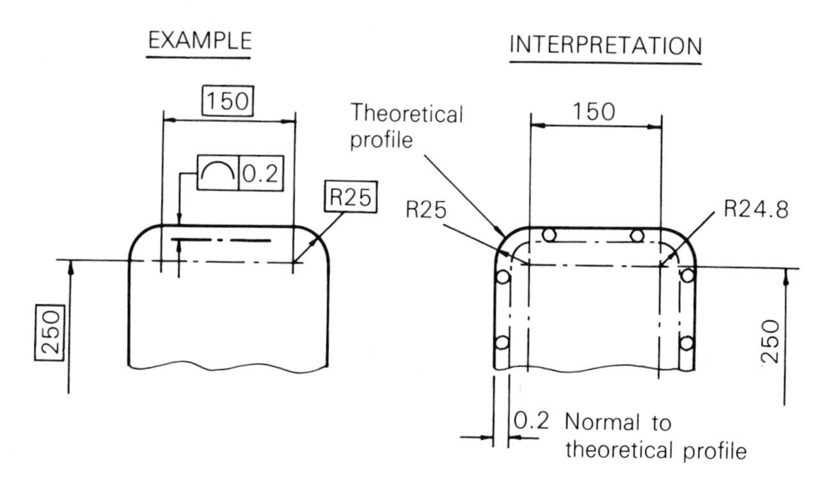

Fig. 5.13 Profile tolerance of a line (with unilateral tolerance zone).

In tolerancing the accuracy of a surface, the theoretical form of the surface is defined by boxed dimensions and a tolerance zone is established in relation to it. The tolerance zone has a constant width normal to the theoretical surface and is the space between two surfaces which envelop spheres of diameter equal to the tolerance value. Fig. 5.14 shows an example of a bilateral tolerance. The curved surface of the part is required to lie between two surfaces which envelop a series of spheres of diameter 0.04 mm with their centres on the surface having the correct geometrical shape.

An equally disposed bilateral zone is implied unless the drawing indicates a unilateral tolerance zone on one side or the other of the theoretical surface.

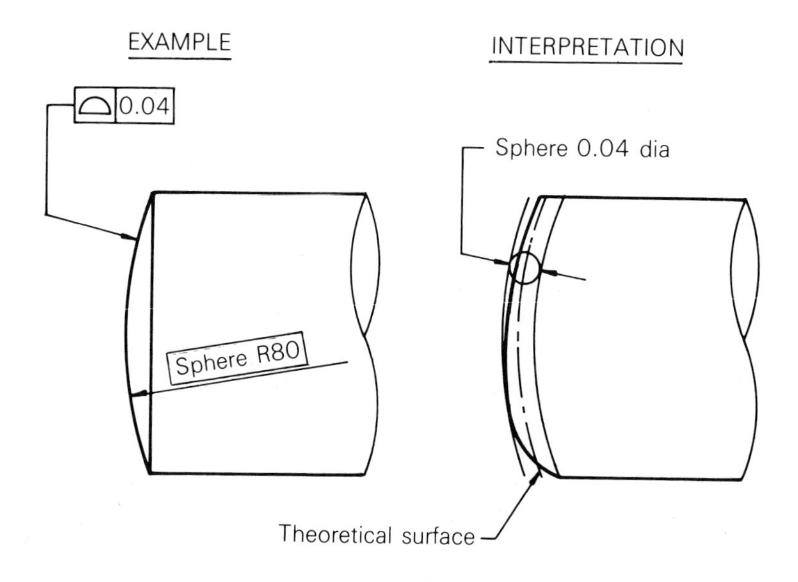

Fig. 5.14 Profile tolerance of a surface (with bilateral tolerance zone).

SYMBOLS FOR TOLERANCING ATTITUDE

In the previous section we examined the tolerances of form which control the deviation of the shape of a feature from its true shape. Additionally a feature may be in error due to lack of parallelism or squareness or incorrect angular position relative to a datum line or surface. These variances are deviations of attitude and the recommended symbols are shown in Fig. 5.15.

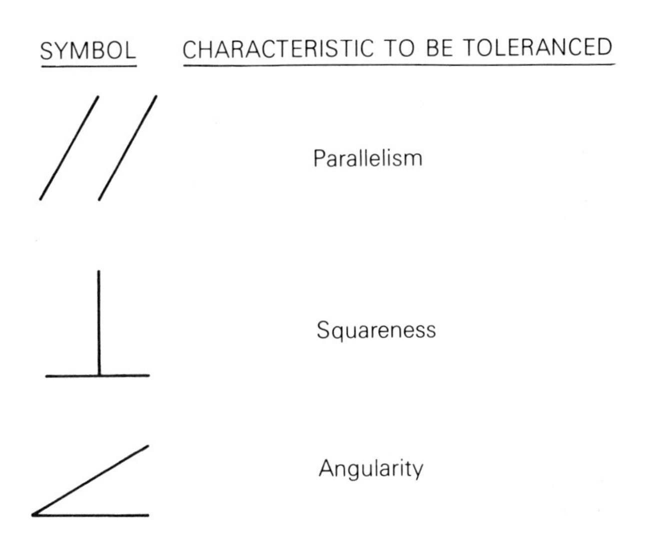

Fig. 5.15 Symbols for tolerancing attitude.

Examples and interpretations of tolerances of parallelism are indicated in Fig. 5.16. The toleranced feature may be a line or a surface and the datum feature may be a line or a plane. The tolerance zone is the area between two parallel lines or the space between two parallel planes which are parallel to the datum feature, and the tolerance value is the distance between the lines or the planes.

In Fig. 5.16a, the axis of the upper cylindrical surface is required to lie between two straight lines, 0.2 mm apart, which are parallel to the datum axis and lie in a vertical plane through it.

In Fig. 5.16b, the surface of the part is required to lie between two parallel planes, 0.2 mm apart, parallel to the datum axis of the hole.

In Fig. 5.16c, the top surface of the part is required to lie between two planes, 0.1 mm apart, parallel to the datum plane.

When a tolerance is stated in respect to squareness, the toleranced feature may be a line or a surface and the datum feature may be a line or a plane. Generally, the tolerance zone is the area between two parallel lines or the space between two parallel lines which are perpendicular to the datum feature, and the tolerance value is the distance between the lines or the planes.

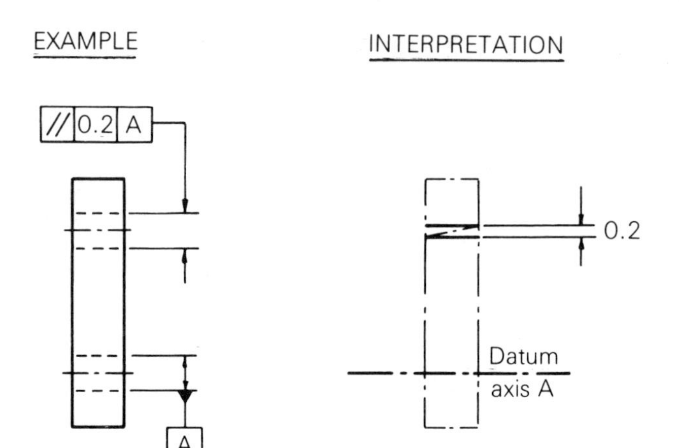

(a) Parallelism of a line with respect to a datum line

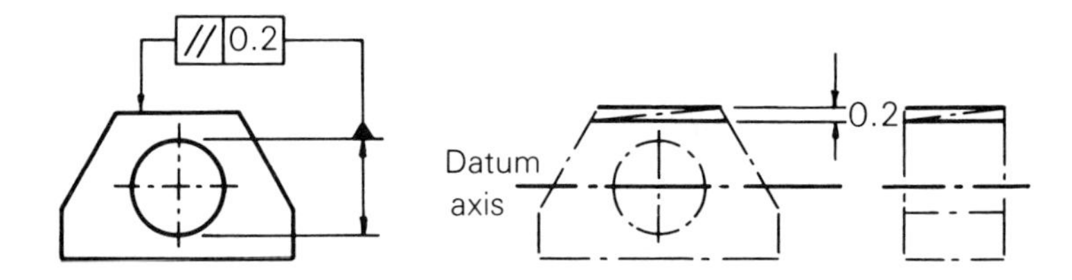

(b) Parallelism of a surface with respect to a datum line

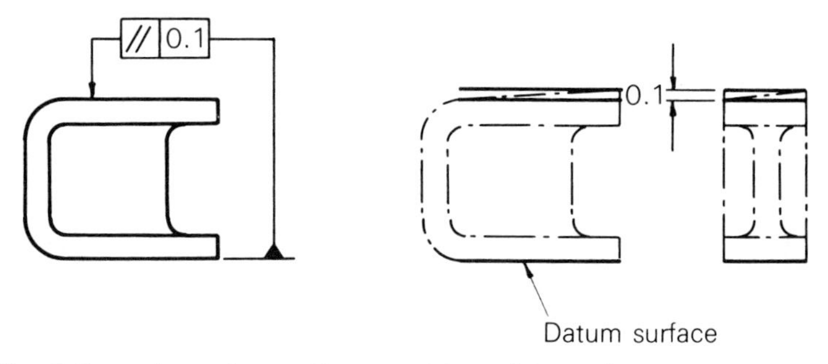

(c) Parallelism of a surface with respect to a datum plane

Fig. 5.16 Tolerances of parallelism.

Fig. 5.17 shows examples and interpretations of tolerances of squareness. In Fig. 5.17a, the vertical face of the part is required to lie between two parallel planes, 0.05 mm apart, which are perpendicular to the datum plane.

In the case of a line with respect to a datum plane (Fig. 5.17b), the tolerance zone may be the space within a cylinder of diameter equal to the tolerance value. In this instance the axis of the cylindrical part is required to be contained within a cylinder of diameter 0.1 mm, the axis of which is perpendicular to the datum plane.

When using a tolerance of angularity, the toleranced feature may be a line or a surface and the datum feature may be a line or a plane. The tolerance zone is the area between two parallel lines which are inclined at a specified angle to a datum feature.

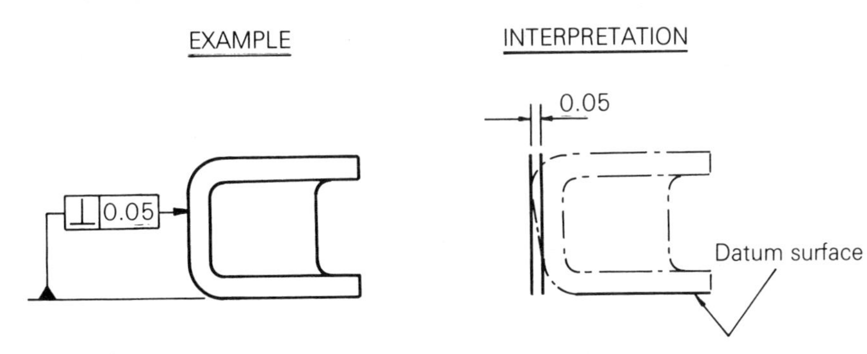

(a) Squareness of a surface with respect to a datum plane

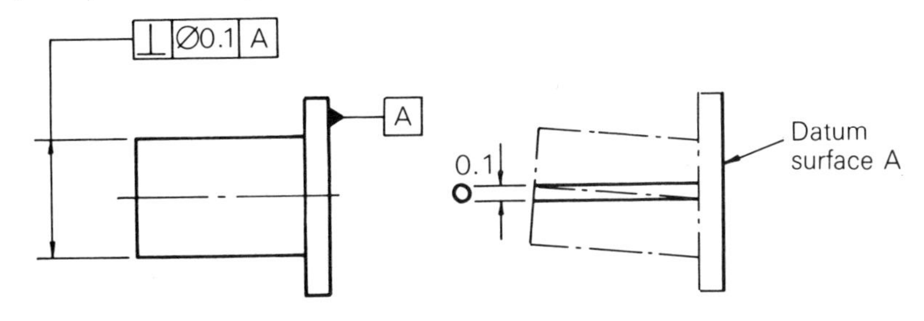

(b) Squareness of a line with respect to a datum plane

Fig. 5.17 Tolerances of squareness.

In Fig. 5.18 angularity of a surface with respect to a datum plane is specified. The inclined surface of the part is required to lie between two planes 0.2 mm apart, which are inclined at 45° to the datum plane.

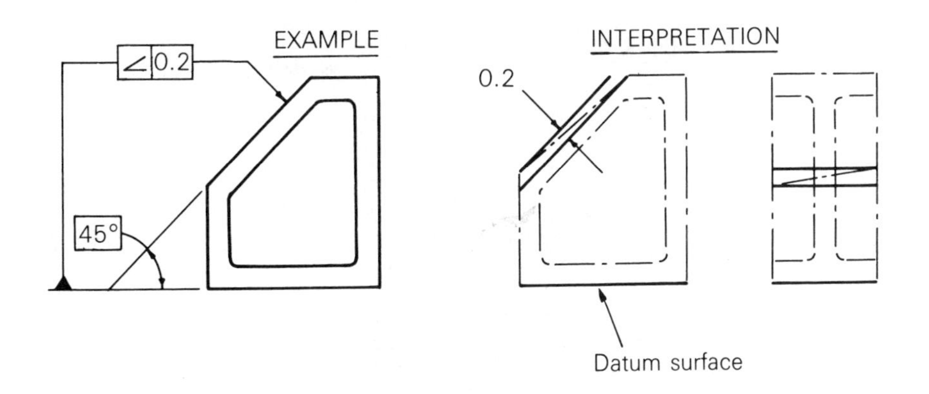

Fig. 5.18 Tolerance of angularity.

SYMBOLS FOR TOLERANCING LOCATION

Tolerances of location control the position of a feature. The symbols for tolerancing location are shown in Fig. 5.19.

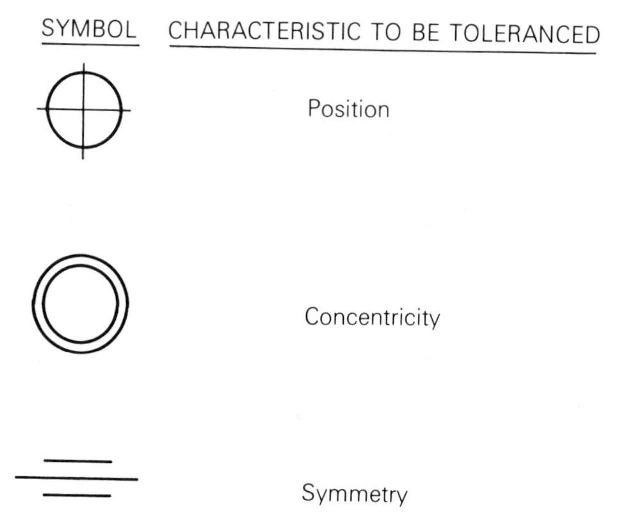

Fig. 5.19 Symbols for tolerancing location.

A tolerance of position limits the deviation of the position of a feature from its specified true position. Examples and interpretations of this type of tolerance are shown in Fig. 5.20.

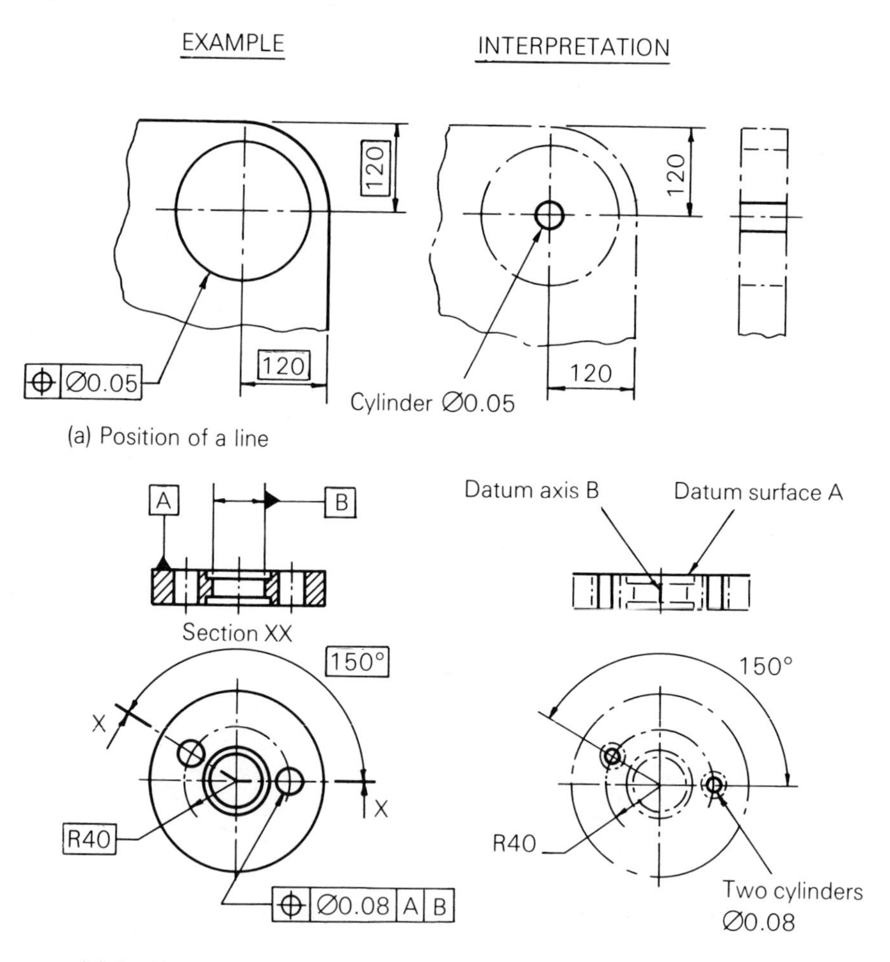

Fig. 5.20 Tolerances of position.

In Fig. 5.20a, the axis of the hole is required to be contained within a cylinder of diameter 0.05 mm with its axis in the specified true position of the axis of the hole. *Note*: these true positions imply also the parallelism and squareness of the tolerance zones with the plane of the drawing.

In Fig. 5.20b, the axes of the two holes are required to be contained within cylinders of diameter 0.08 mm with their axes in the specified true positions of the holes in relation to the datum surface A, and the axis of the central hole B.

Tolerances of concentricity are shown in Fig. 5.21. A concentricity tolerance is a particular case of a positional tolerance in which the toleranced feature and the datum feature are circles or cylinders. The tolerance limits the deviation of the position of the centre of the toleranced feature from its true position. The tolerance value is the diameter of the tolerance zone. In the example shown, the axes of the three smaller cylindrical parts are required to be contained within one cylinder of diameter 0.03 mm.

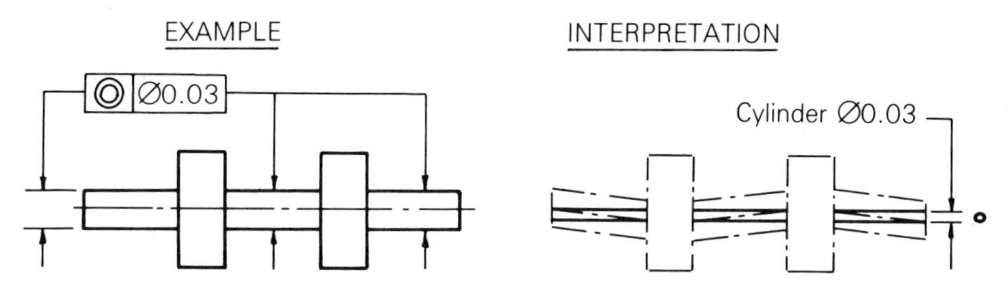

Fig. 5.21 Tolerances of concentricity.

Examples and interpretations of tolerances of symmetry are given in Fig. 5.22. A symmetry tolerance is a particular case of a positional tolerance in which the position of the feature is specified by its symmetrical relationship to a datum.

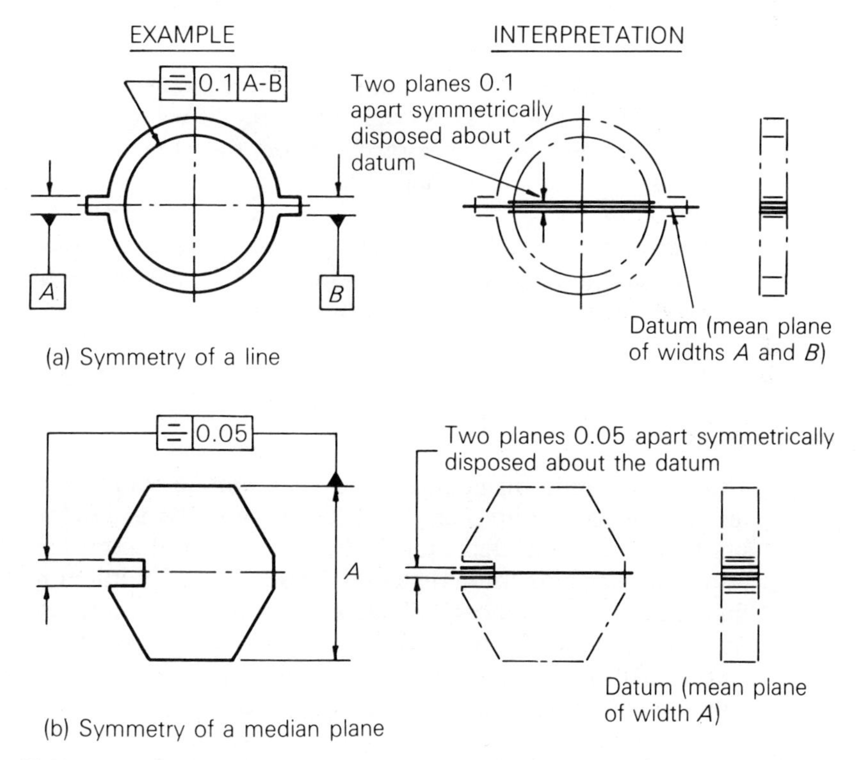

Fig. 5.22 Tolerances of symmetry.

Fig. 5.22a gives an example of symmetry of a line. The axis of the hole is required to lie between two parallel planes, 0.1 mm apart, which are symmetrically disposed about the common median plane of the rectangular protrusions.

Fig. 5.22b gives an example of symmetry of a median plane. The median plane of the slot is required to lie between two parallel planes, 0.05 mm apart, which are symmetrically disposed about the median plane of the width A.

TYPES OF FIT

Earlier in the chapter it was stated that a component cannot be manufactured to an exact size. It is, however, the primary function of a production engineer to arrange to make components with regular precision, because, in the manufacture of a machine, jig or fixture, quality is of foremost importance. Precision is the condition of accuracy necessary to guarantee the operation of a component as intended by the designer. Mating surfaces of components in an assembly need to be manufactured with precision so that they can operate effectively, but because it is impossible to produce a component to an exact size, some variation must be allowed in production.

This allowable variation, which produces the means of accomplishing the required precision, is known as the *tolerance* of the part. Consider a part with a nominal size of 20 mm diameter. The part may be satisfactory if the diameter is larger or smaller by, say, 0.02 mm; then:

$$\text{the largest acceptable diameter is } 20+0.02 = 20.02 \text{ mm dia}$$

$$\text{the smallest acceptable diameter is } 20-0.02 = 19.98 \text{ mm dia}$$

$$\text{the tolerance} = 20.02-19.98 = 0.04 \text{ mm}$$

The largest acceptable diameter is known as the *upper limit* and the smallest as the *lower limit*.

In order to achieve an exact function there has to be a relationship between mating parts of an assembly. If a single assembly is being made, the type of fit necessary may be decided in discussion between the designer and the toolroom fitter or machinist. It may be sufficient to quote running fit, sliding fit, drive fit or press fit and leave the actual sizes to the judgement of the fitter or machinist. If, say, 1000 assemblies are being manufactured, however, interchangeability (any part A must fit any part B) will be necessary and the allowance of the fit and the tolerance of each part must be specified more satisfactorily.

There are three classes of fit which apply to mating components, as follows.

(a) Clearance Fit

This is the condition in which the male part (a shaft or protrusion) is smaller than the female part (a hole or slot). This type of fit is illustrated in Fig. 5.23a where a shaft, of maximum diameter 49.97 mm and minimum diameter 49.94 mm, is to fit into a hole of maximum diameter 50.05 mm and minimum diameter 50.00 mm. The shaft and hole tolerances will allow many different variations in the type of clearance fit. The tightest fit will be obtained with the largest shaft (49.97 mm diameter) mating with the smallest hole (50.00 mm diameter) to give a minimum clearance of 0.03 mm. The loosest fit will be obtained with the smallest shaft (49.94 mm diameter) mating with the largest hole (50.05 mm diameter) to give a maximum clearance of 0.11 mm.

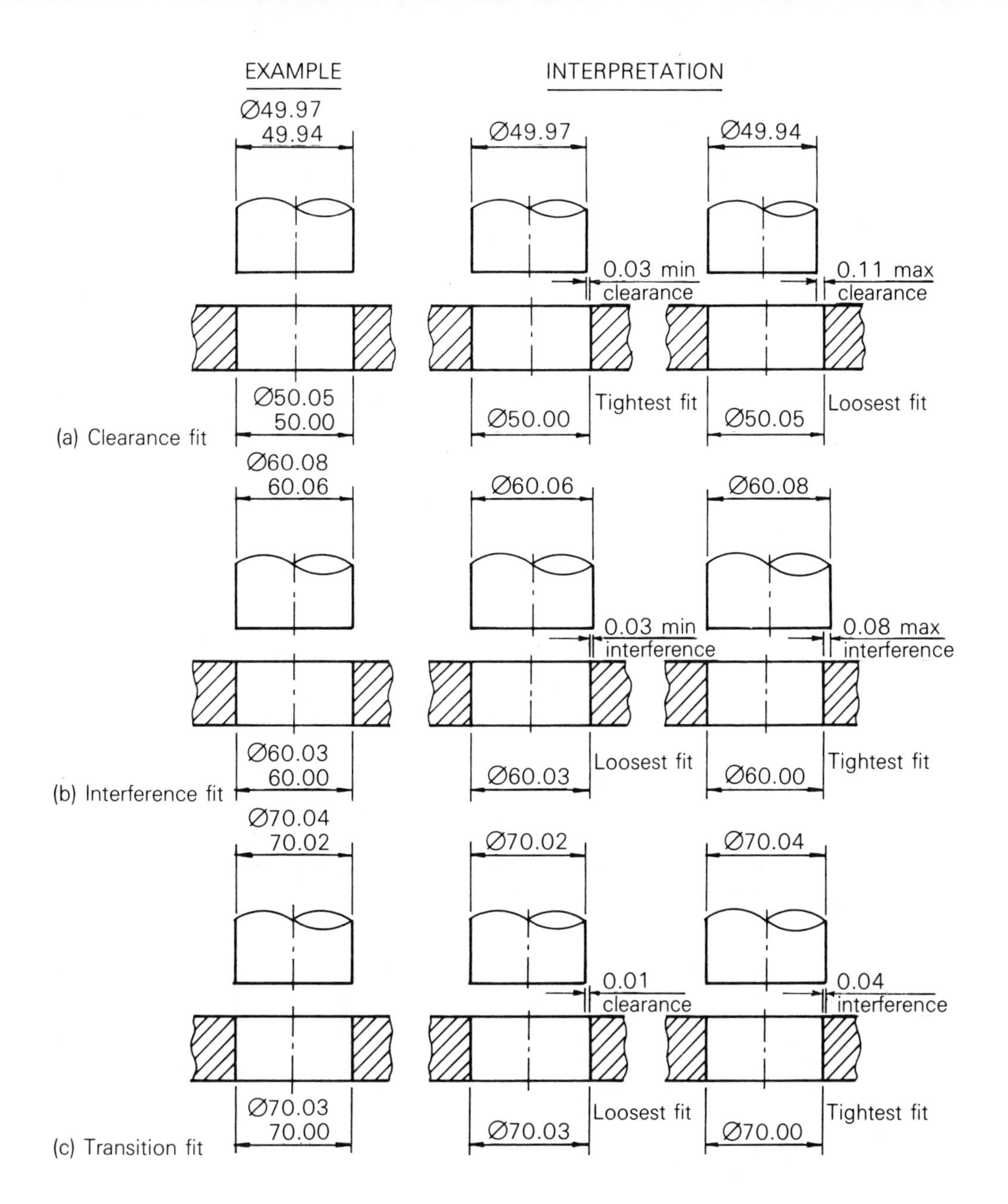

Fig. 5.23 Types of fit.

(b) Interference Fit

This is the condition in which the male part is always larger than the female part and is illustrated in Fig. 5.23b. A shaft, of maximum diameter 60.08 mm and minimum diameter 60.06 mm, is to fit into a hole of maximum diameter 60.03 mm and minimum diameter 60.00 mm. The loosest fit will be obtained with the smallest shaft (60.06 mm diameter) mating with the largest hole (60.03 mm diameter) to give a minimum interference of 0.03 mm. The tightest fit will be obtained with the largest shaft (60.08 mm diameter) mating with the smallest hole (60.00 mm diameter) to give a maximum interference of 0.08 mm.

(c) Transition Fit

This is the condition in which either a clearance or an interference fit may be achieved. This type of fit is illustrated in Fig. 5.23c, where a shaft, of maximum diameter 70.04 mm and minimum diameter 70.02 mm, is to fit into a hole of maximum diameter 70.03 mm and minimum diameter 70.00 mm. A clearance fit will be obtained when the smallest shaft (70.02 mm diameter) mates with the largest hole (70.03 mm diameter) and an interference fit will be obtained when the largest shaft (70.04 mm diameter) mates with the smallest hole (70.00 mm diameter).

The ISO system (BS 4500) is designed to provide a comprehensive range of limits and fits for engineering purposes. It is based on a series of tolerances graded to suit all classes of work from the finest to the coarsest. BS 4500A — see Table 5.1 — gives a small selection of fits from the system which will satisfy most normal requirements.

The position of a tolerance zone relative to the zero line of basic size controls the type of fit which is to be achieved. For any mating system the position of the tolerance zone for the hole is designated by a capital letter and that of the shaft by a small letter. BS 4500A uses only H holes, where the lower limit is the basic size, and shafts c to h, k, n, p and s, and covers a range of fits suitable for most engineering applications. The size of the tolerance is controlled by a number — the smaller the number the finer the tolerance. Fig. 5.24 illustrates the principles of the system.

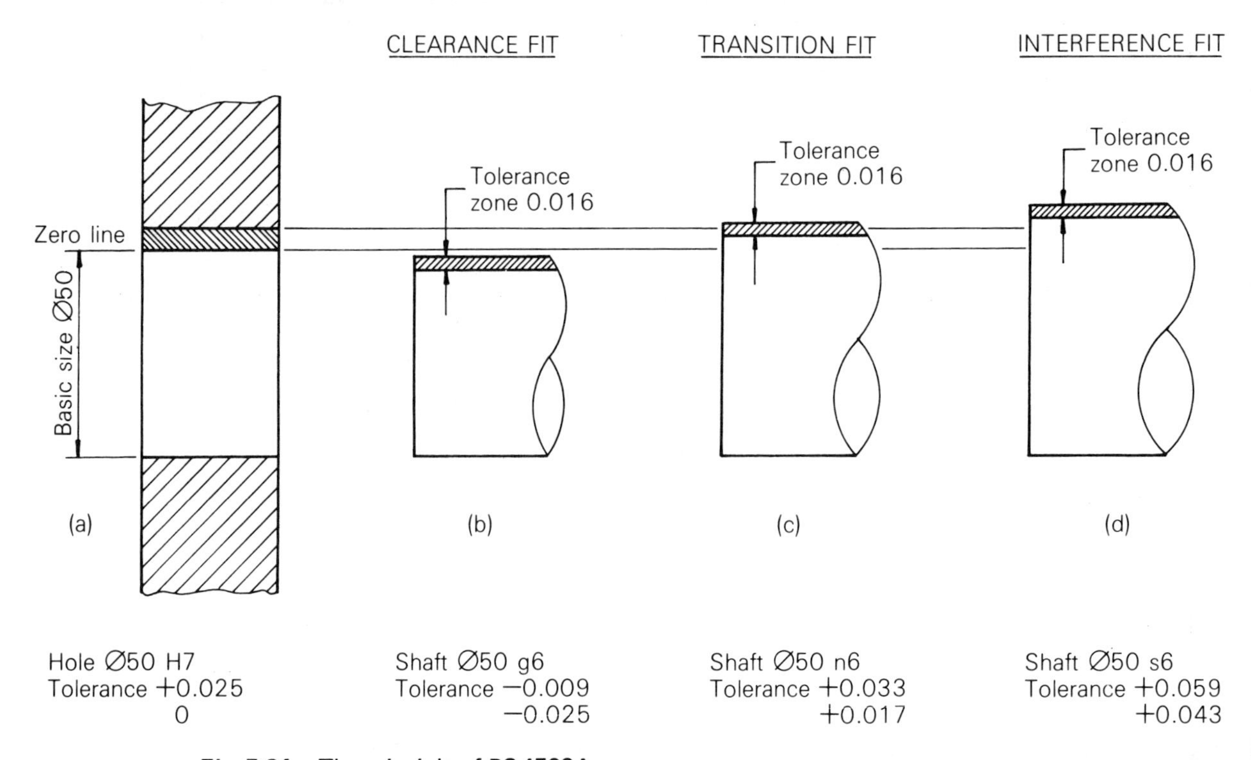

Fig. 5.24 The principle of BS 4500A.

Fig. 5.24a shows a hole of basic size 50 mm diameter and classification H7. From BS 4500A the limits of tolerance are seen to be

$$\begin{matrix} +0.025 \\ 0 \end{matrix}\ \text{mm}$$

In Figs 5.24b, c and d, shafts have been selected with the same number classification, i.e. 6, and consequently have equal tolerance zones, i.e. 0.016 mm.

Fig. 5.24b shows a shaft with classification g6. The limits of tolerance are

$$\begin{matrix} -0.009 \\ -0.025 \end{matrix} \text{mm}$$

When mating with a hole H7, a clearance fit would be achieved, and would be designated ∅ 50 H7/g6.

Fig. 5.24c shows a shaft with a classification n6. The limits of tolerance are

$$\begin{matrix} +0.033 \\ +0.017 \end{matrix} \text{mm}$$

When mating with a hole H7, a transition fit would be achieved and would be designated ∅ 50 H7/n6.

Fig. 5.24d shows a shaft with classification s6. The limits of tolerance are

$$\begin{matrix} +0.059 \\ +0.043 \end{matrix} \text{mm}$$

When mating with a hole H7, an interference fit would be achieved and would be designated ∅ 50 H7/s6.

Problem: Define the maximum and minimum limits of size, for the hole and shaft, in the following mating systems:

(a) 55 mm H8/f7
(b) 225 mm H7/p6
(c) 7.5 mm H7/k6.

Which kind of fit is achieved in each instance?

Solution: (a) Hole ∅ 55 H8:

Tolerance $\begin{matrix} +0.046 \\ 0 \end{matrix}$ mm
Maximum limit 55.046 mm
Minimum limit 55.000 mm

Shaft ∅ 55 f7:

Tolerance $\begin{matrix} -0.030 \\ -0.060 \end{matrix}$ mm
Maximum limit 54.970 mm
Minimum limit 54.940 mm

The shaft would always be smaller than the hole and therefore a *clearance fit* is achieved.

(b) Hole ∅ 225 H7:

Tolerance $\begin{matrix} +0.046 \\ 0 \end{matrix}$ mm
Maximum limit 225.046 mm
Minimum limit 225.000 mm

Shaft ∅ 225 p6:

Tolerance $\begin{matrix} +0.079 \\ +0.050 \end{matrix}$ mm
Maximum limit 225.079 mm
Minimum limit 225.050 mm

The shaft would always be bigger than the hole and therefore an *interference fit* would be achieved.

Extracted from BS 4500: 1969

Table 5.1 *British Standard, selected ISO fits — hole basis*

Nominal sizes		Clearance fits									
Diagram to scale for 25 mm. diameter		H11 / c11		H9 / d10		H9 / e9		H8 / f7		H7 / g6	
		Tolerance		Tolerance		Tolerance		Tolerance		Tolerance	
Over	**To**	**H11**	**c11**	**H9**	**d10**	**H9**	**e9**	**H8**	**f7**	**H7**	**g6**
mm	mm	0.001 mm	0.001 mm	0.001 mm	0.001 mm	0.001 mm	0.001 mm	0.001 mm	0.001 mm	0.001 mm	0.001 mm
—	3	+60 0	−60 −120	+25 0	−20 −60	+25 0	−14 −39	+14 0	−6 −16	+10 0	−2 −8
3	6	+75 0	−70 −145	+30 0	−30 −78	+30 0	−20 −50	+18 0	−10 −22	+12 0	−4 −12
6	10	+90 0	−80 −170	+36 0	−40 −98	+36 0	−25 −61	+22 0	−13 −28	+15 0	−5 −14
10	18	+110 0	−95 −205	+43 0	−50 −120	+43 0	−32 −75	+27 0	−16 −34	+18 0	−6 −17
18	30	+130 0	−110 −240	+52 0	−65 −149	+52 0	−40 −92	+33 0	−20 −41	+21 0	−7 −20
30	40	+160 0	−120 −280	+62 0	−80 −180	+62 0	−50 −112	+39 0	−25 −50	+25 0	−9 −25
40	50	+160 0	−130 −290								
50	65	+190 0	−140 −330	+74 0	−100 −220	+74 0	−60 −134	+46 0	−30 −60	+30 0	−10 −29
65	80	+190 0	−150 −340								
80	100	+220 0	−170 −390	+87 0	−120 −260	+87 0	−72 −159	+54 0	−36 −71	+35 0	−12 −34
100	120	+220 0	−180 −400								
120	140	+250 0	−200 −450	+100 0	−145 −305	+100 0	−84 −185	+63 0	−43 −83	+40 0	−14 −39
140	160	+250 0	−210 −460								
160	180	+250 0	−230 −480								
180	200	+290 0	−240 −530	+115 0	−170 −355	+115 0	−100 −215	+72 0	−50 −96	+46 0	−15 −44
200	225	+290 0	−260 −550								
225	250	+290 0	−280 −570								
250	280	+320 0	−300 −620	+130 0	−190 −400	+130 0	−110 −240	+81 0	−56 −108	+52 0	−17 −49
280	315	+320 0	−330 −650								
315	355	+360 0	−360 −720	+140 0	−210 −440	+140 0	−125 −265	+89 0	−62 −119	+57 0	−18 −54
355	400	+360 0	−400 −760								
400	450	+400 0	−440 −840	+155 0	−230 −480	+155 0	−135 −290	+97 0	−68 −131	+63 0	−20 −60
450	500	+400 0	−480 −880								

		Transition fits				Interference fits					
H7 / h6		H7 / k6		H7 / n6		H7 / p6		H7 / s6		Holes / Shafts	
Tolerance		Tolerance		Tolerance		Tolerance		Tolerance		Nominal sizes	
H7	h6	H7	k6	H7	n6	H7	p6	H7	s6	Over	To
0.001 mm	0.001 mm	0.001 mm	0.001 mm	0.001 mm	0.001 mm	0.001 mm	0.001 mm	0.001 mm	0.001 mm	mm	mm
+10 0	−6 0	+10 0	+6 0	+10 0	+10 +4	+10 0	+12 +6	+10 0	+20 +14	—	3
+12 0	−8 0	+12 0	+9 +1	+12 0	+16 +8	+12 0	+20 +12	+12 0	+27 +19	3	6
+15 0	−9 0	+15 0	+10 +1	+15 0	+19 +10	+15 0	+24 +15	+15 0	+32 +23	6	10
+18 0	−11 0	+18 0	+12 +1	+18 0	+23 +12	+18 0	+29 +18	+18 0	+39 +28	10	18
+21 0	−13 0	+21 0	+15 +2	+21 0	+28 +15	+21 0	+35 +22	+21 0	+48 +35	18	30
+25 0	−16 0	+25 0	+18 +2	+25 0	+33 +17	+25 0	+42 +26	+25 0	+59 +43	30	40
										40	50
+30 0	−19 0	+30 0	+21 +2	+30 0	+39 +20	+30 0	+51 +32	+30 0	+72 +53	50	65
								+30 0	+78 +59	65	80
+35 0	−22 0	+35 0	+25 +3	+35 0	+45 +23	+35 0	+59 +37	+35 0	+93 +71	80	100
								+35 0	+101 +79	100	120
+40 0	−25 0	+40 0	+28 +3	+40 0	+52 +27	+40 0	+68 +43	+40 0	+117 +92	120	140
								+40 0	+125 +100	140	160
								+40 0	+133 +108	160	180
+46 0	−29 0	+46 0	+33 +4	+46 0	+60 +31	+46 0	+79 +50	+46 0	+151 +122	180	200
								+46 0	+159 +130	200	225
								+46 0	+169 +140	225	250
+52 0	−32 0	+52 0	+36 +4	+52 0	+66 +34	+52 0	+88 +56	+52 0	+190 +158	250	280
								+52 0	+202 +170	280	315
+57 0	−36 0	+57 0	+40 +4	+57 0	+73 +37	+57 0	+88 +62	+57 0	+226 +190	315	355
								+57 0	+244 +208	355	400
+63 0	−40 0	+63 0	+45 +5	+63 0	+80 +40	+63 0	+108 +68	+63 0	+272 +232	400	450
								+63 0	+292 +252	450	500

BRITISH STANDARDS INSTITUTION, 2 Park Street, London, W1A 2BS
SBN: 580 05766 6

(c) Hole ∅ 7.5 H7:

Tolerance $^{+0.015}_{0}$ mm
Maximum limit 7.515 mm
Minimum limit 7.500 mm

Shaft ∅ 7.5 k6:

Tolerance $^{+0.010}_{+0.001}$ mm
Maximum limit 7.510 mm
Minimum limit 7.501 mm

The shaft may or may not be bigger than the hole and therefore a *transition fit* would be achieved.

Application to Practical Problems

The student needs to be able to relate the type of fit—clearance, transition or interference—with the functional requirements of a mating assembly. Table 5.2 gives a general guide to the type of fit which can be achieved by the various combinations in BS 4500A together with the appropriate class of work which would be used to accomplish them successfully.

Table 5.2 *Types of fit obtained from BS 4500A*

Mating Assembly	*Type of Fit* — *BS 4500 Class*	*Type of Fit* — *Interpretation*	*Class of Work*
H11/c11	Clearance	Very loose running	Drilling, rough turning
H9/d10		Loose running	Boring/milling
H9/e9		Easy running	Boring/batch lathe work
H8/f7		Medium running	Reaming/good-quality turning
H7/g6		Close running	Careful reaming/grinding
H7/h6		Sliding	
H7/k6	Transition	Easy push	
H7/n6		Push	
H7/p6	Interference	Light press	
H7/s6		Press	

Examples of Mating Fits

(a) Light press fit, H7/p6 (Fig. 5.25):
a flanged bearing located into a bearing housing.

(b) Medium running fit, H8/f7 (Fig. 5.25):
good-quality fit obtained between bearing and rotating spindle.

(c) Press fit, H7/s6 (Fig. 5.26):
used as a semi-permanent assembly where a guide pin is secured into a moving bolster.

(d) Sliding fit, H7/h6 (Fig. 5.26):
used for location purposes; the guide pin moves in and out of the lower bolster plate during operation.

(e) Easy push fit H7/k6 (Fig. 5.27):

used to locate a part where movement is not required; in this instance a machine vice would be positioned on a machine table and a tenon used to locate it in a table tee-slot.

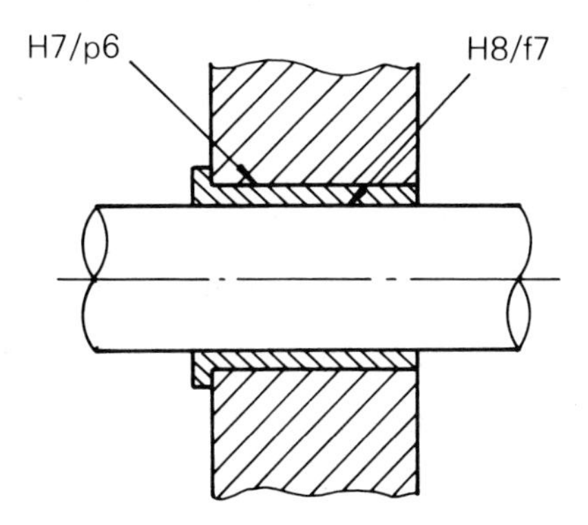

Fig. 5.25

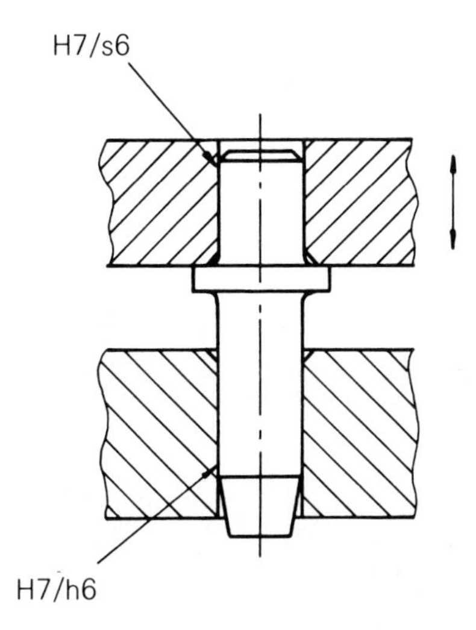

Fig. 5.26

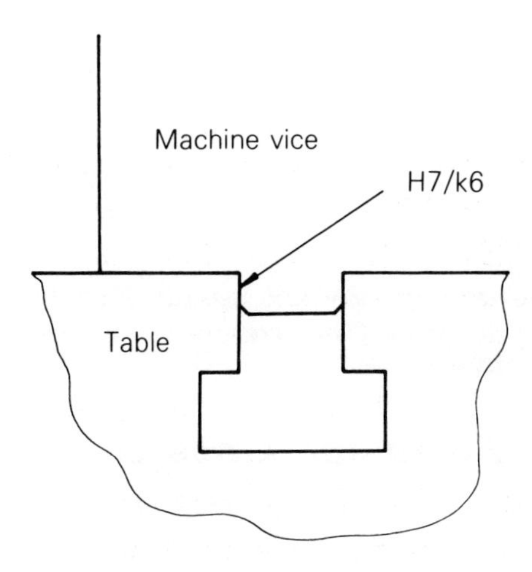

Fig. 5.27

DIMENSIONING DRAWINGS

The student will have had some experience of dimensioning drawings from previous studies in Engineering Drawing I. The dimensioning of a drawing will conclude its presentation, therefore the utmost attention must be given to this facet because it is a vital part of draughtsmanship, and, unless it is done correctly, it will spoil even the most proficient drawing.

Principles of Dimensioning

A dimension for a particular feature should appear once only on the drawing.

There should be no more dimensions than are necessary to define a component.

Dimensions relating to a single feature should be placed adjacent to it in a single view rather than spread out over several views.

The dimensions which directly affect the function of a component should be stated on the drawing. Fig. 5.28 shows the functional and non-functional dimensions of an assembly. Dimensions marked F directly influence the compatibility of the two components on assembly. The non-functional dimensions, NF, are chosen to suit production purposes.

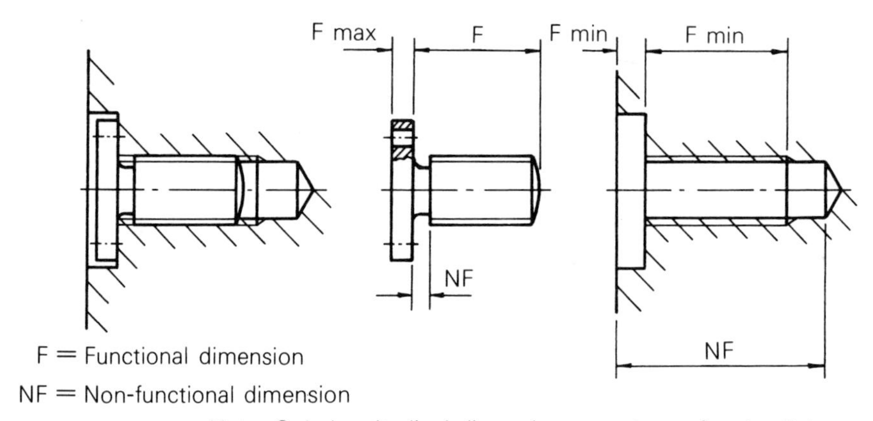

Fig. 5.28 Functional dimensions.

Whenever possible, standard sizes should be used, particularly for holes, screw threads, nuts, bolts, studs, pins, etc., and for work where the sizes and surface finish of standard stock, such as bright drawn or centreless ground material, would be practicable.

Production processes are not usually stated on drawings unless they are essential for a particular function. Drill sizes, however, are often quoted.

Projection Lines, Dimension Lines and Leaders

Projection lines are continuous thin lines, 0.3 mm wide (see Table 2, BS 308: Part 1), which project from just clear of the outline of a component and extend to a little beyond the dimension line—see Fig. 5.29.

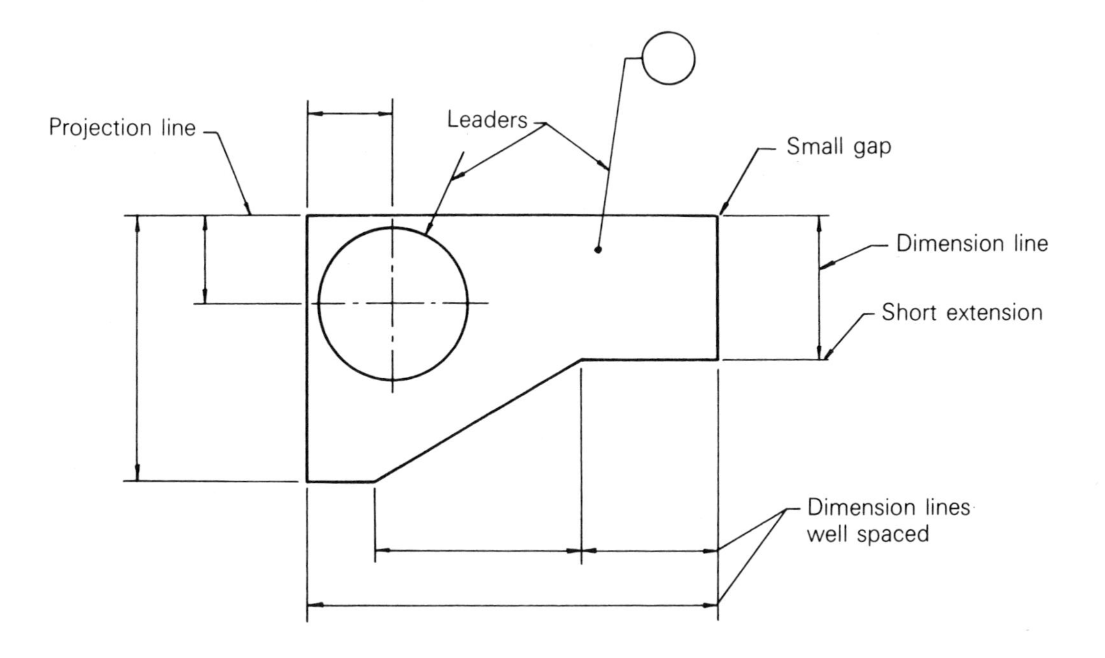

Fig. 5.29 Projection lines, dimension lines and leaders.

Where projection lines refer to points on a surface or at an intersection, they should touch the points. For clarification, these points are often emphasised by small dots—see Fig. 5.30.

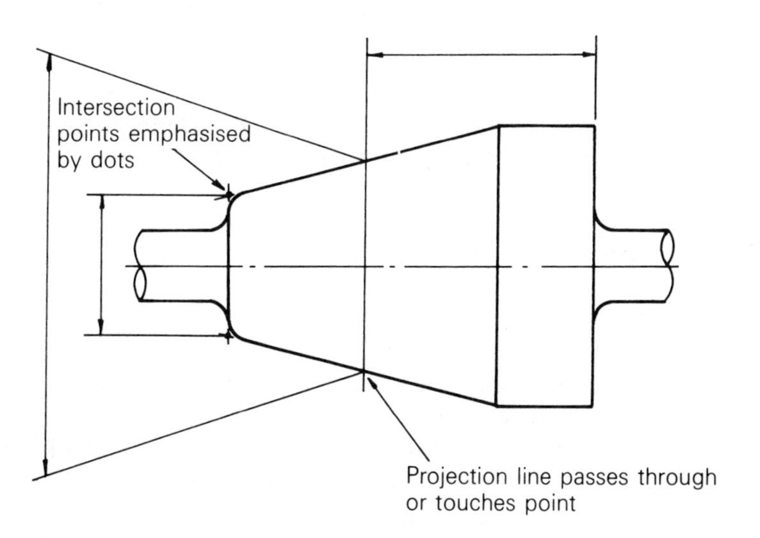

Fig. 5.30 Projection lines from points of intersection.

Dimension lines are also continuous thin lines; they should be placed outside the outline of the component wherever practicable and should touch the projection lines (Fig. 5.29).

Dimension lines and projection lines should not cross other lines unless it is unavoidable.

Dimension lines should be placed on the view which shows the relevant features most clearly.

Arrowheads should be not less than 3 mm long and should touch the projection line. Illustrations of good and bad arrowheads are shown in Fig. 5.31.

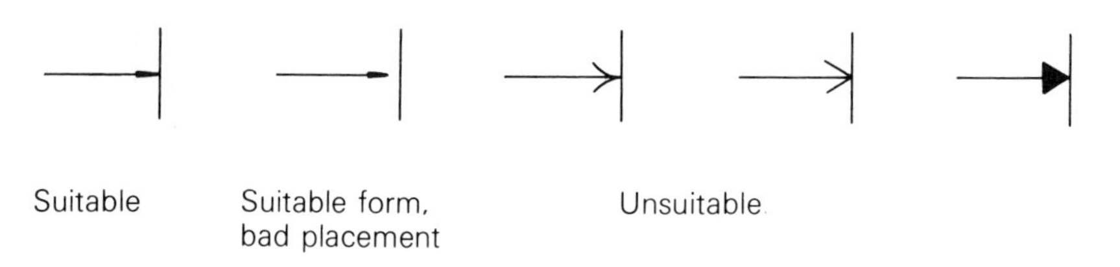

Fig. 5.31 Arrowheads.

Leaders are used to indicate where dimensions or notes are intended to apply and are terminated by arrowheads or dots. Arrowheads should always terminate on a line but dots should be within the outline of a component (Fig. 5.29). A leader line is also a continuous thin line.

Dimensions

Drawings should state the unit of dimension, which generally should be expressed in millimetres on metric engineering drawings. In some cases it may be more appropriate to use metres or micrometres. Where drawings are dimensioned in one unit, the unit abbreviation may be omitted provided the drawing carries a statement of the unit used.

A point is used for the decimal marker on engineering drawings and should be placed on the base line, for example, 47.62. Dimensions of less than unity should be preceded by a zero, for example, 0.78.

Angular dimensions should be expressed in degrees, minutes and seconds, for example, 15°25′30″. When an angle is less than one degree it should be preceded by 0°, for example, 0°30′.

Figures and letters used in dimensioning should be large enough for easy reading. The minimum character height, for A2 and A4 drawing sheets, is 2.5 mm (full details of minimum character heights are shown in Table 3, BS 308: Part 1.) To help in reading dimensions, figures should be placed so that they can be read from the bottom or from the right of the drawing.

Larger dimensions should be placed outside smaller dimensions. Dimensions which are not drawn to scale should be underlined.

Where an overall dimension is shown, one of the intermediate distances is redundant and should not be dimensioned. In some cases, however, all the intermediate dimensions need to be shown and consequently the overall dimension is added as an 'auxiliary' dimension. An auxiliary dimension is shown in parentheses (...).

Fig. 5.32 shows the arrangement of dimensions.

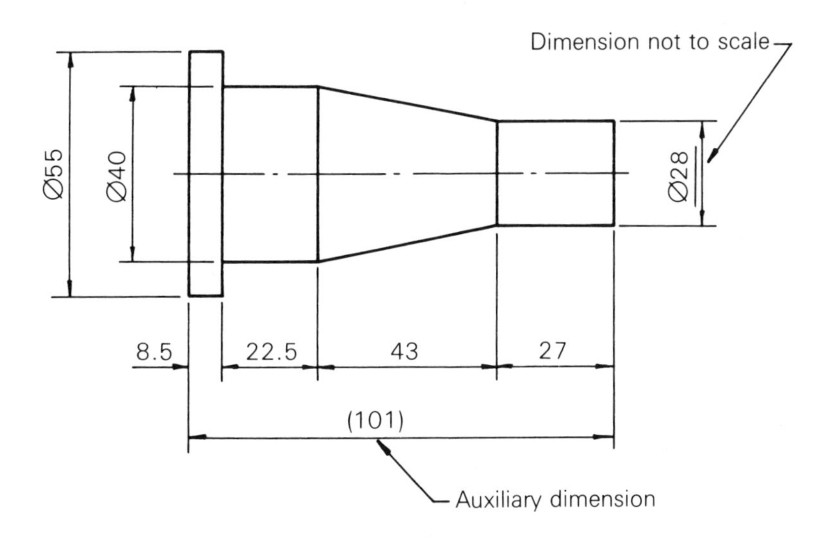

Fig. 5.32 Arrangement of dimensions.

Toleranced Dimensions

Where it is necessary to tolerance an individual dimension, one of the methods shown in Fig. 5.33 should be used.

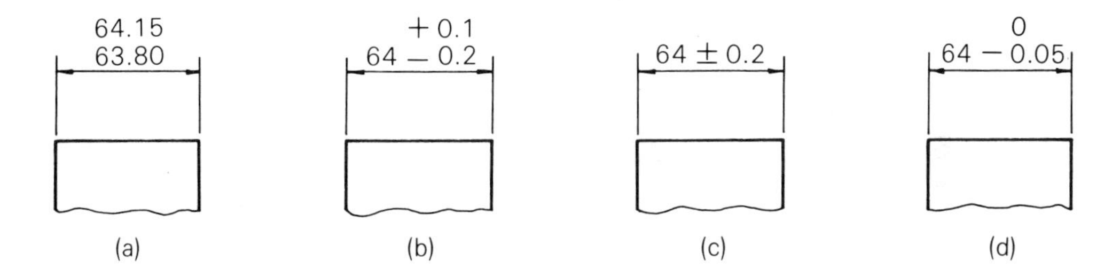

Fig. 5.33 Dimensioning limits of size.

Fig. 5.33a specifies the upper and lower limits of size. This is a highly satisfactory method because it obviates the need for the machinist to undertake any calculations, and eliminates a consequential risk of error.

Figs. 5.33b, c and d show alternative systems by specifying a size with limits of tolerance above and below that size. Both limits of tolerance should be expressed to the same number of decimal places except where one of these limits is zero, when it should be expressed by the figure '0'.

Where fits are taken from BS 4500A, the appropriate symbols should be expressed as shown in Fig. 5.34. For a shaft, with a nominal size of 64 mm diameter and a tolerance n6, any of the methods shown in Figs 5.34a, b or c may be used. For a hole, with the same basic diameter and a tolerance H7, the methods shown in Figs 5.34d, e or f may be used.

If these dimensions become numerous, the values of the designations may be placed in tabulated form on the drawing. In such cases the indication on the dimension would be as specified in Fig. 5.34a, for shafts, or Fig. 5.34d for holes.

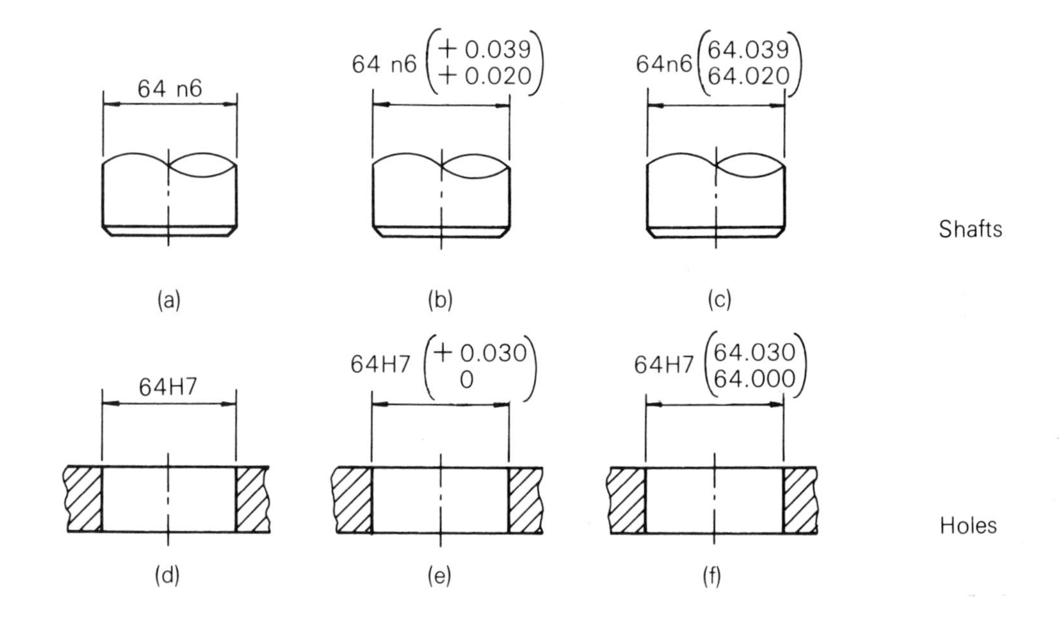

Fig. 5.34 Dimensioning limits and fits.

The recommendations for the tolerancing of angular dimensions are shown in Fig. 5.35.

30° max
30°30′ 30°0′
90° ± 2°
45°30′ ± 0°2′30″
$15^{\circ}\,{}^{0}_{-1^{\circ}}$

Fig. 5.35 Tolerancing of angular dimensions.

A general tolerance note may be added to a drawing, if appropriate:

TOLERANCE EXCEPT WHERE OTHERWISE STATED ± . . .

Where it is necessary to specify only one limit of size of a dimension, the recommendations shown in Fig. 5.36 should be used; Fig. 5.36a shows a minimum length of full thread required and Fig. 5.36b shows a maximum radius.

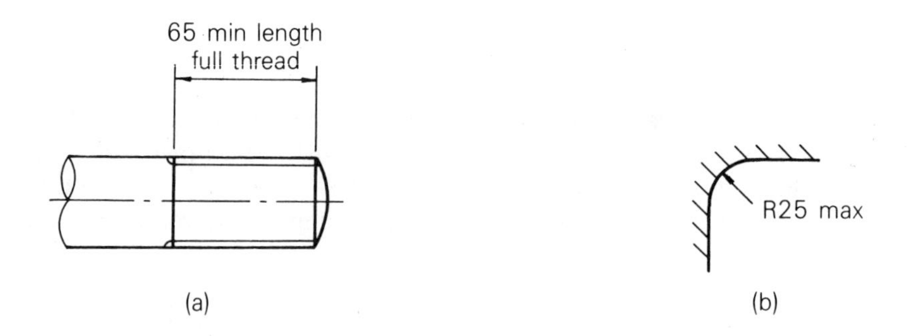

Fig. 5.36 Single limits of size.

Methods of Dimensioning Common Features

Diameters

When dimensioning a diameter, the size should be preceded by the symbol ∅. Dimensions of diameters should be placed on the most appropriate view for clarity—see Fig. 5.37. If the dimensions had been added to the end-view in this instance, it would have led to confusion.

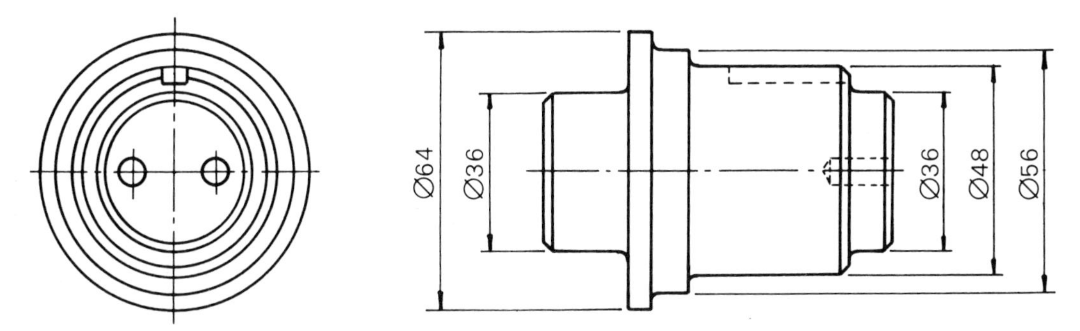

Fig. 5.37 Dimensioning diameters.

Circles may be dimensioned by any of the methods shown in Fig. 5.38.

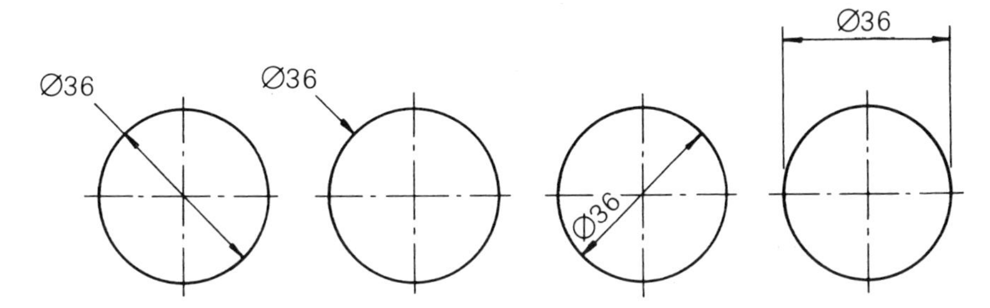

Fig. 5.38 Dimensioning circles.

Radii

Radii may be dimensioned by any of the methods illustrated in Fig. 5.39. The abbreviation R should precede the dimension. The dimension line should pass through, or be in line with, the centre of the arc.

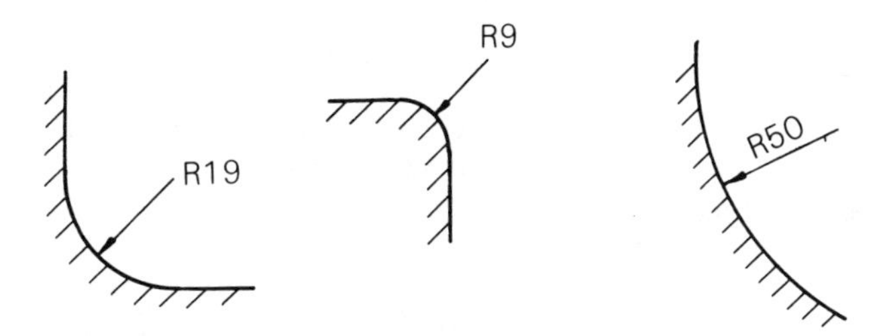

Fig. 5.39 Dimensioning radii.

Size and position of holes

When dimensioning holes, the method of production may be added (drill, ream, etc.). The positions of holes may be defined by spacing them on circles as shown in Fig. 5.40a. Alternatively they may be defined by their rectangular co-ordinate dimensions as shown in Fig. 5.40b.

Methods of dimensioning countersinks and counterbores are also shown in Fig. 5.40.

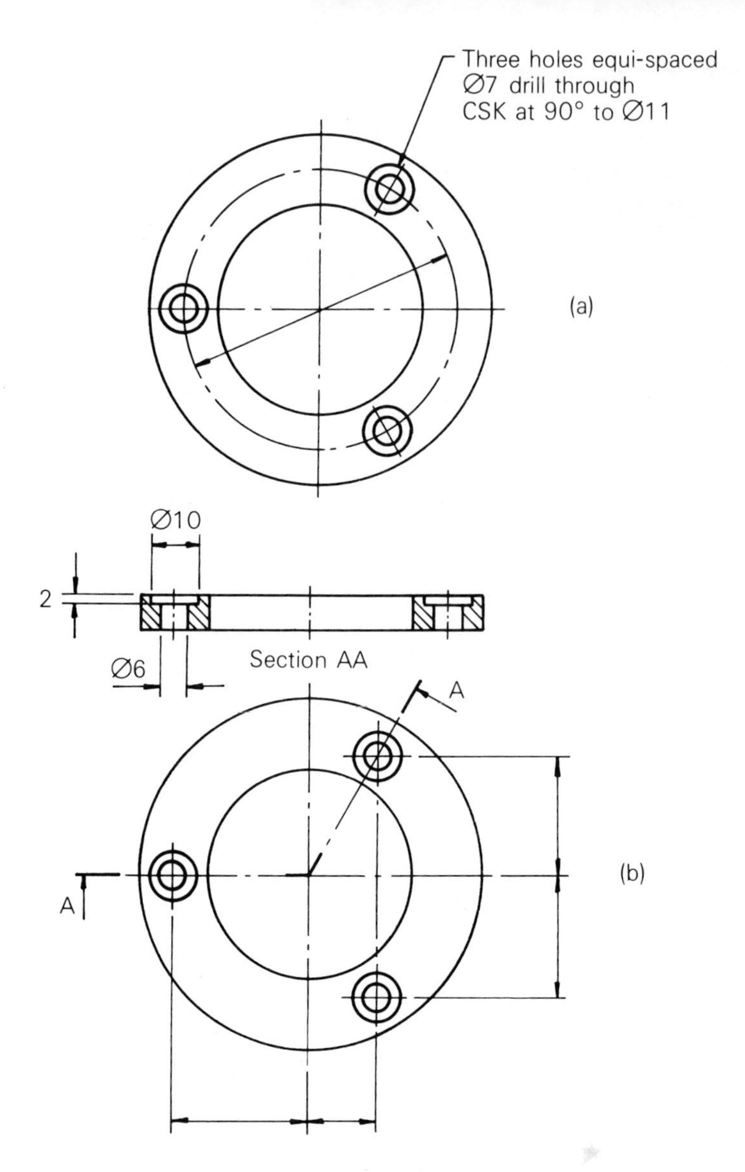

Fig. 5.40 Dimensioning sizes and positions of holes.

Chamfers

Most chamfers are angled at 45°. Methods of dimensioning 45° chamfers are shown in Fig. 5.41. The practice of specifying a chamfer by a note and a leader should not be encouraged.

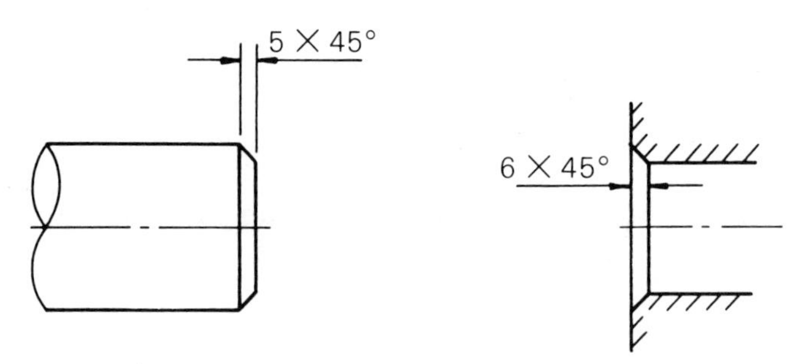

Fig. 5.41 Chamfers at 45°

Screw threads

Fig. 5.42a illustrates the dimensioning of a screw thread. The designation means that the thread is a metric one, of outside diameter 12 mm and pitch 1.75 mm. 6g denotes the accuracy of the thread; the letter g indicates an external thread.

The dimensioning of threaded holes is shown in Fig. 5.42b. One hole is to be threaded all the way through the component and the other is to be threaded to a depth of 15 mm minimum. The designation is similar to that for screw threads, the letter H indicating an internal thread.

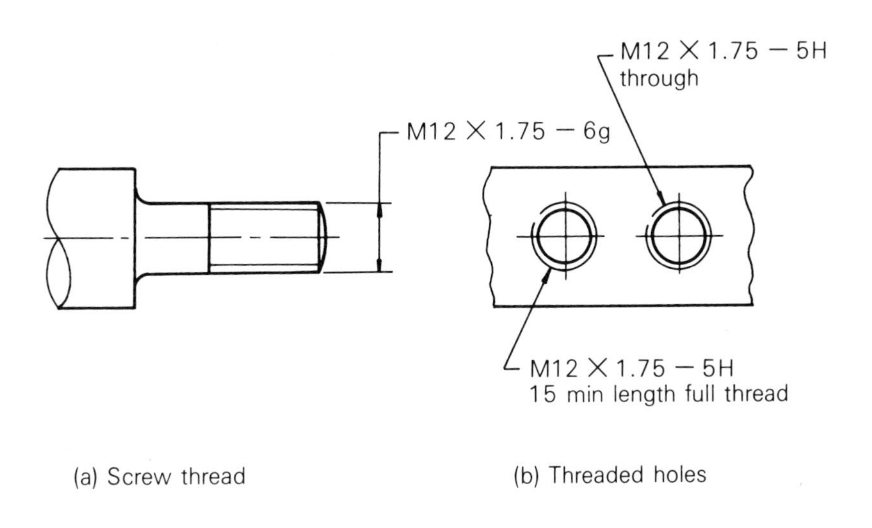

Fig. 5.42 Dimensioning screw threads and threaded holes.

Keyways

Recommended methods of dimensioning keyways in shafts and hubs are shown in Fig. 5.43.

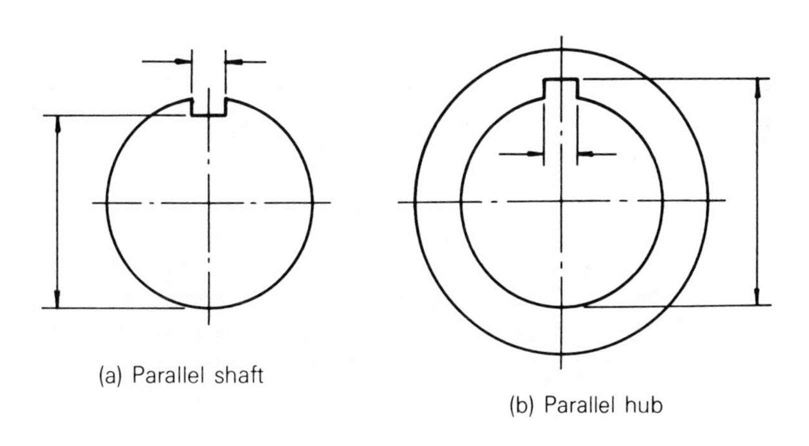

Fig. 5.43 Dimensioning keyways.

Diamensioning from a datum

Two principal systems are used for linear dimensioning:

(a) chain dimensioning
(b) dimensioning from a common datum.

Fig. 5.44 illustrates both methods on a common component.

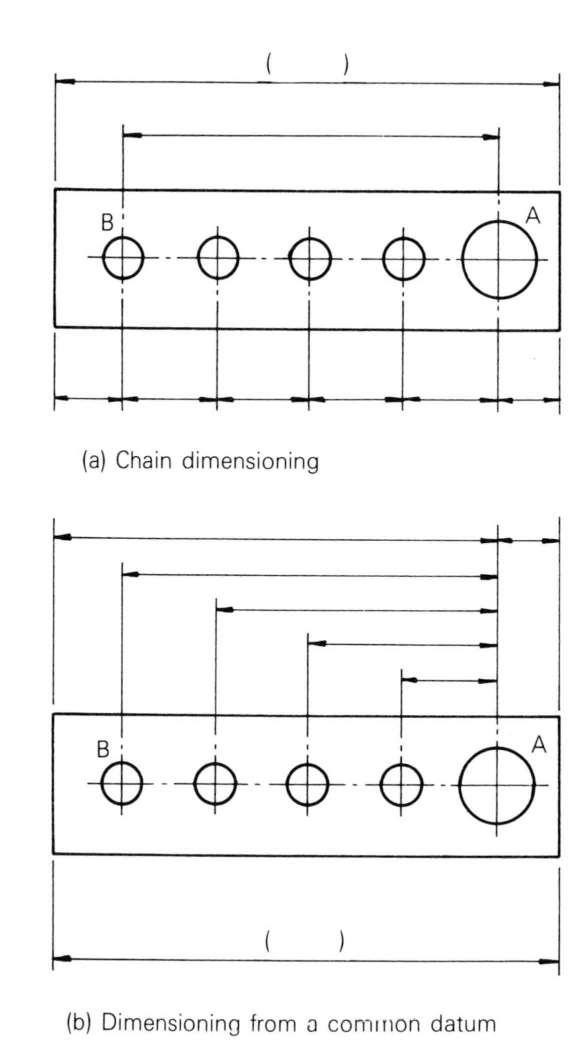

Fig. 5.44 Chain and datum dimensioning.

Chain dimensioning should only be used where the possible cumulative error does not affect the functional requirements of the part. If an overall tolerance of ± 0.1 mm is stated on the drawing (Fig. 5.44a), a cumulative error, between holes A and B, of ± 0.4 mm may occur when using a chain dimensioning system.

If the centre of hole A is used as a datum, as shown in Fig. 5.44b, errors will not accumulate because each dimension would be subject to the general tolerance and the error between holes A and B will be ± 0.1 mm maximum.

Tapered Features

Several methods are used to dimension tapered components—two methods are illustrated in Fig. 5.45.

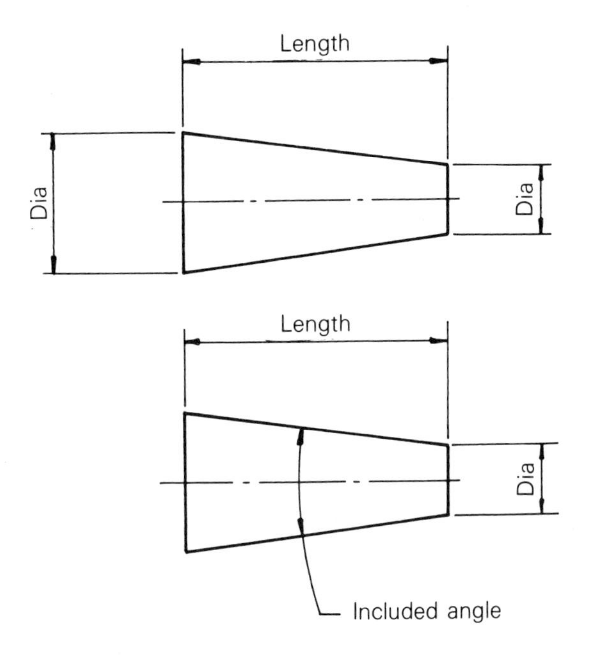

Fig. 5.45 Dimensioning tapered components.

There should be no more dimensions than are necessary to define the tapered feature although additional dimensions for information purposes may be given as auxiliary dimensions.

This section has given the salient principles of dimensioning and the methods of applying tolerances of size on engineering drawings as recommended in BS 308: Part 2: 1972. The student should now examine this Standard in detail to become familiar with all the recommendations.

Examples of the use of BS 308: Part 2 and Part 3 in dimensioning and tolerancing working drawings are shown in Chapter 6.

SELF-TEST 5

Select the correct option(s). *Note*: There may be more than one correct answer.

1. A geometrical tolerance of a feature controls the maximum permissible overall variation of:

(a) form
(b) position
(c) size
(d) attitude.

2. Fig. 5.46 shows the drawing of a component. Fig. 5.47 shows the sizes of the component obtained by inspection. The errors are of:

(a) size and location
(b) size and attitude
(c) location and attitude
(d) size, location and attitude.

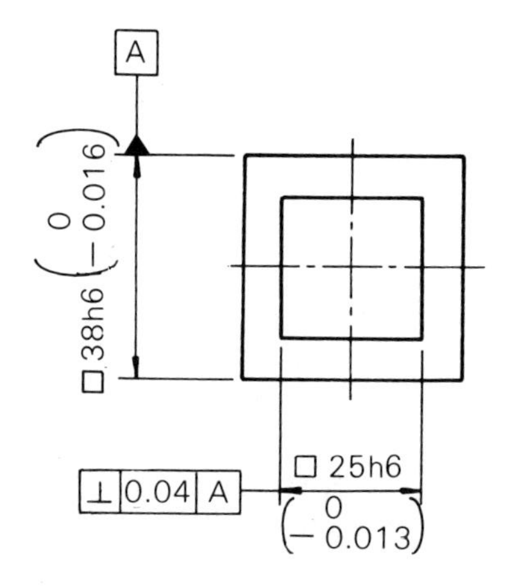

Fig. 5.46

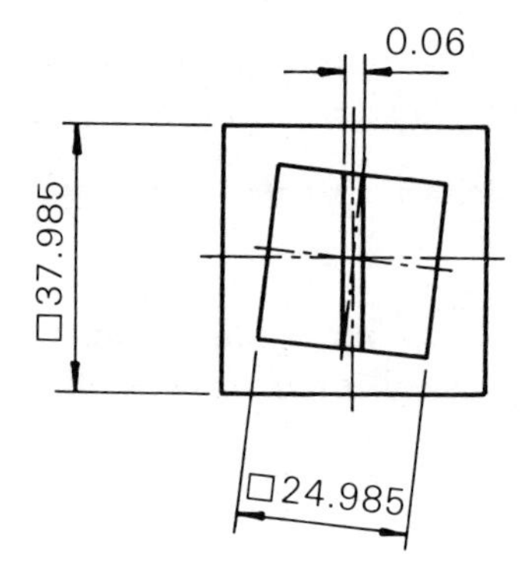

Fig. 5.47

3. The geometrical tolerance indicated in Fig. 5.48 means that the feature which must lie within a tolerance of 0.05 is its:

(a) roundness
(b) straightness
(c) cylindricity
(d) surface profile.

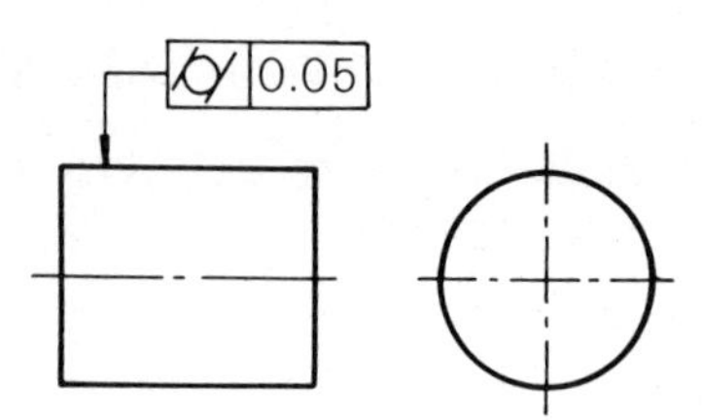

Fig. 5.48

4.

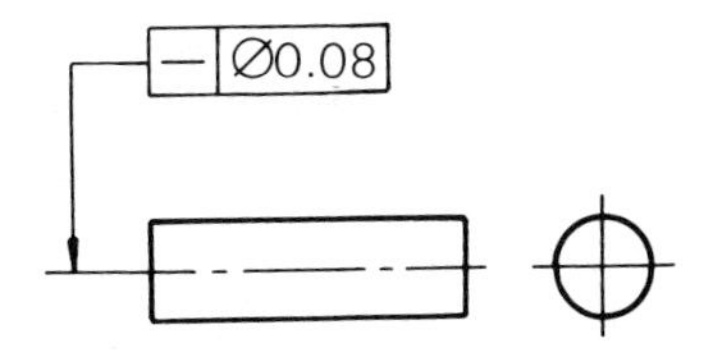

Fig. 5.49

Fig. 5.49 shows a tolerance of straightness applied to a spindle. Four manufactured components are inspected and the sizes noted. The component(s) which is (are) outside the tolerance is (are):

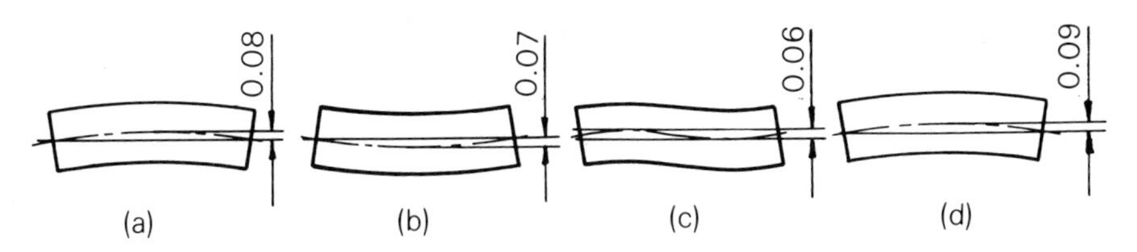

5. Fig. 5.50 shows an example of tolerance of position of a point. It means that the point is required to lie within:

(a) 0.4 mm of the position stated
(b) 0.4 mm of the stated diameter of the point
(c) a circle of diameter 0.4 mm
(d) 28 ± 0.4 mm of the edges of the component.

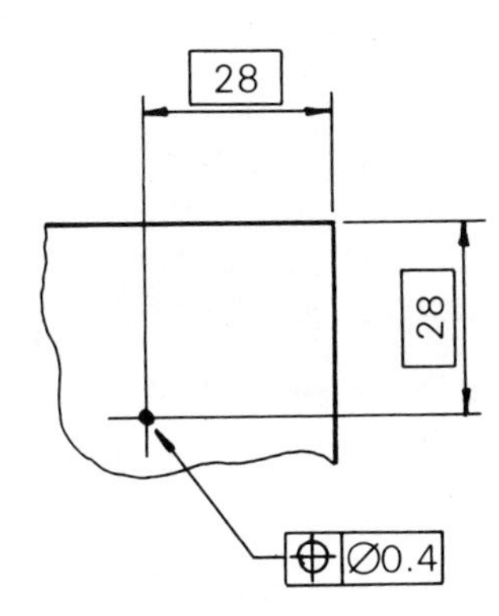

Fig. 5.50

6. A hole tolerance is expressed as:

$$70\,\text{H7}\begin{pmatrix}+0.030\\0\end{pmatrix}$$

The tolerances of three shafts are expressed as:

$$\text{I}\quad 70\,\text{g6}\begin{pmatrix}-0.010\\-0.029\end{pmatrix}\qquad \text{II}\quad 70\,\text{h6}\begin{pmatrix}0\\-0.019\end{pmatrix}\qquad \text{III}\quad 70\,\text{k6}\begin{pmatrix}+0.021\\+0.002\end{pmatrix}$$

The resulting respective fits are:

(a) clearance, transition, interference
(b) clearance, clearance, transition
(c) clearance, transition, transition
(d) clearance, interference, interference.

7. Four different methods of dimensioning circles are shown in Fig. 5.51. The methods recommended by BS 308: Part 2: 1972 are:

(a) I and II only
(b) I, II and III only
(c) II and IV only
(d) II, III and IV only.

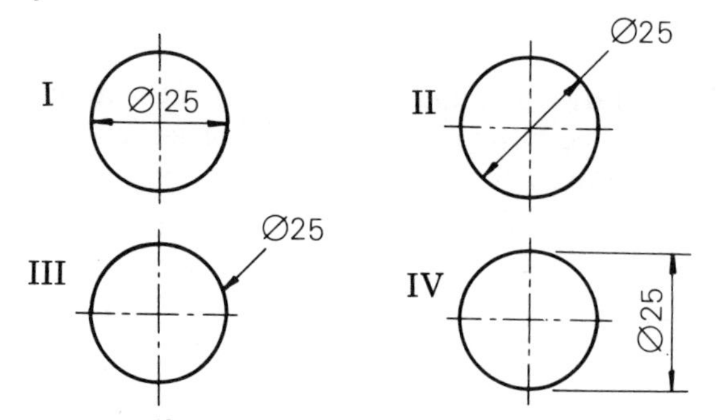

Fig. 5.51

8. A thread is designated M6 × 1. This means the thread is metric and has a nominal diameter of:

(a) 6 mm and a pitch of 1 mm
(b) 1 mm and a pitch of 6 mm
(c) 6 mm and a length of 1 mm
(d) 1 mm and a length of 6 mm.

9. The respective classes of fit for: I, a phosphor bronze bush in a housing; II, an idler gear on a shaft; and III, sliding and rotary motion between two mating parts, are:

(a) interference, clearance, clearance
(b) interference, transition, clearance
(c) transition, transition, clearance
(d) clearance, interference, interference.

10. Fig. 5.52 includes six linear dimensions. The number of incorrectly drawn linear dimensions is:

(a) one
(b) two
(c) three
(d) four.

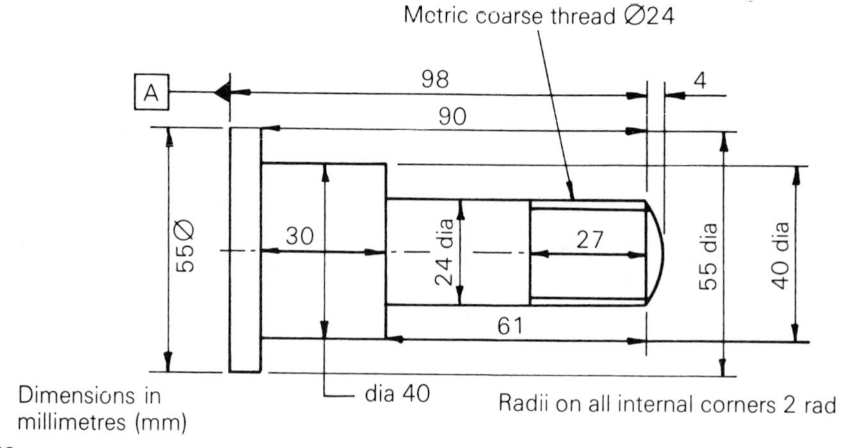

Fig. 5.52

EXERCISE 5

1. A fit is given as ∅116 H7/g6. Explain what is meant by this classification and state the class of fit. Determine the maximum and minimum limits of size of the mating components.

2. Give a practical application of the mating assembly detailed in Question 1.

3. Three classes of fit apply to mating components. Draw a diagram to show how the position of the tolerance zone defines each class.

4. For each of the following hole and shaft combinations state the class of fit:

	Hole	*Shaft*
(a)	$210^{+0.046}_{0}$	$210^{+0.033}_{+0.004}$
(b)	$120^{+0.054}_{0}$	$120^{-0.036}_{-0.071}$
(c)	$2^{+0.010}_{0}$	$2^{+0.006}_{0}$
(d)	$3^{+0.010}_{0}$	$3^{+0.020}_{+0.014}$
(e)	$316^{+0.057}_{0}$	$316^{+0.226}_{+0.190}$

5. For each of the holes and shafts in Question 4 determine the maximum and minimum limits of size.

6. Fig. 5.52 shows a component which is drawn or dimensioned incorrectly. Re-draw the component to the sizes given and dimension it to the recommendations given in BS 308: Part 2: 1972.

7. Pair-up the correct symbol with the characteristic to be toleranced in Fig. 5.53.

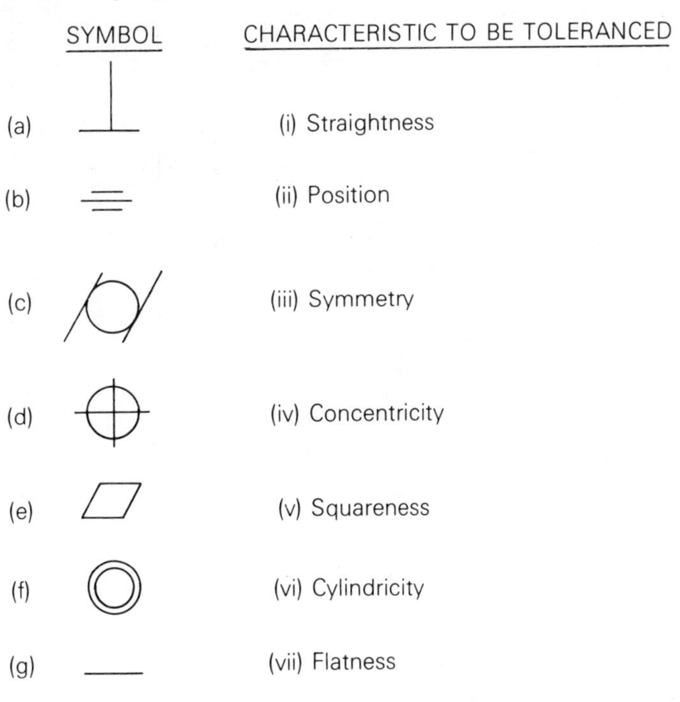

Fig. 5.53

8. Re-draw the component shown in Fig. 5.54, replacing the notes with the dimensioning and tolerancing systems recommended in BS 308: Part 2 and Part 3: 1972.

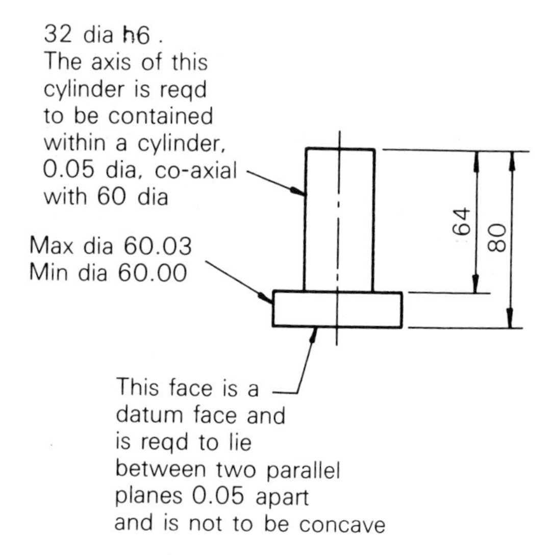

Fig. 5.54

9. With the aid of a sketch, show how the axis of a hole, which is required to lie between two parallel planes 0.1 mm apart when inclined at 45° to the axis of a cylindrical component, should be defined by the system recommended in BS 308: Part 3: 1972.

10. Determine the meaning of the screw thread designation and each geometrical tolerance in Fig. 5.55.

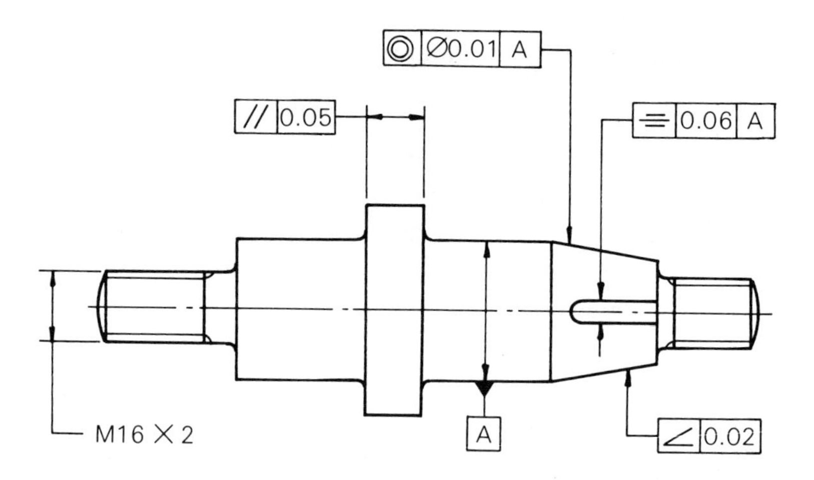

Fig. 5.55

SECTION F

ENGINEERING DRAWING

Production of Working Drawings

After reading this chapter you should be able to:

* ★ produce working engineering drawings to British Standard Specification (G);
* ★ complete a title block to British Standard Specification (S);
* ★ select a suitable scale for a given component (British Standard Specification) (S);
* ★ draw auxiliary views of given components (S);
* ★ produce working drawings from given engineering components conforming to British Standard Specification, including tolerancing to BS 308: Part 3: 1972 (ISO) (S);
* ★ produce working drawings of engineering assemblies incorporating hidden detail and sections and conforming to British Standard Specification (for example, tail stock, valve, jig, machine vice) (S);
* ★ sketch given engineering components and assemblies in order to produce working drawings (S);
* ★ state, when given details of the use of a component, which of the common engineering materials, for example, iron, aluminium alloys, copper-based alloys, is likely to be used and state which mechanical properties of the material make it appropriate for that use (S);
* ★ write parts lists to British Standard Specification on working drawings (S);
* ★ produce fully dimensioned drawings from given assembly drawings (S).

(G) = general TEC objective
(S) = specific TEC objective

ENGINEERING MATERIALS

A designer will consider many factors before deciding which material to use for the manufacture of a component:

Is the mass critical?
Does it need to have good thermal or electrical conductivity?
Is the operating temperature within the critical range of some materials?
Will expansion or contraction arise?
Does the component need to have anti-corrosive properties?

When giving thought to these factors the designer will consider the physical and chemical properties of the alternative materials.

The mechanical properties of materials will also be considered. These include:

Mechanical property	*Ability*
Hardness	Will withstand wear and indentation.
Strength	Will resist fracture when subject to a force.
Fatigue resistance	Will withstand repetitive or cyclic loading (such as in a car suspension unit).
Toughness	Will withstand impact loading.
Brittleness	This is the opposite of toughness; will not withstand an impact load.
Ductility	Will deform considerably under load before breaking.
Malleability	Will compress without fracture.
Elasticity	Will return to its original shape after being deformed by a load.

The designer will also consider manufacturing quantity. The process used to produce one, one hundred or ten thousand components will vary and consequently the type of material used in alternative processes may also be different.

In general the types of materials used in engineering can be split into two categories: *metallic* and *non-metallic* materials. Metallic materials can also be sub-divided into ferrous metals (those containing iron) and non-ferrous metals (such as copper, aluminium and their alloys) —see Fig. 6.1.

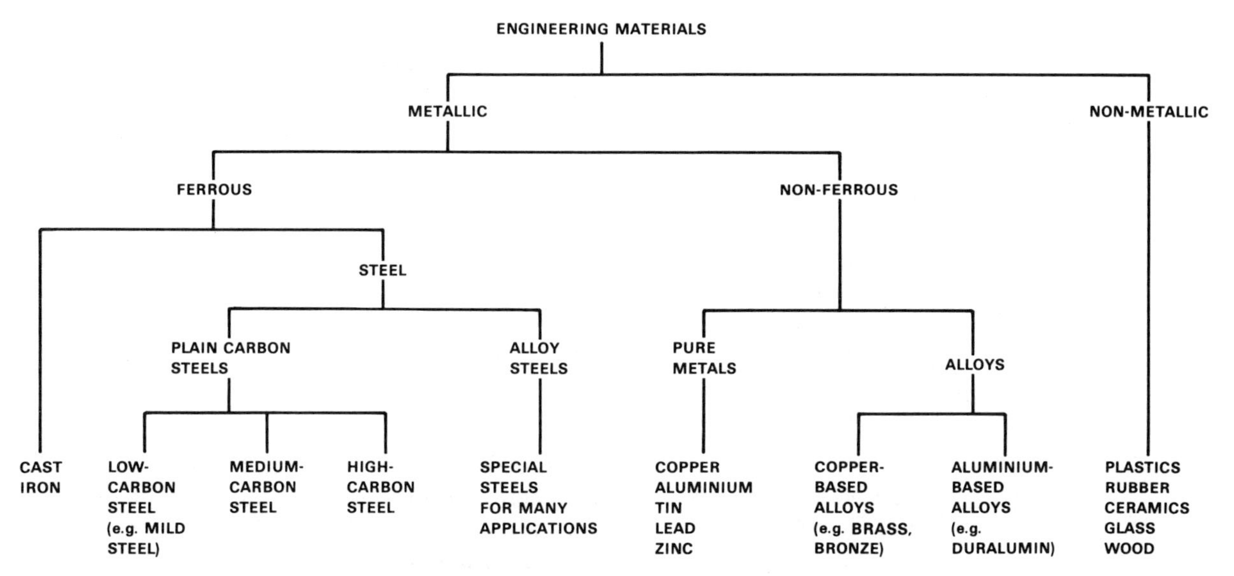

Fig. 6.1 General engineering materials.

Properties and Uses of Common Engineering Materials

Ferrous metals

Type	*Property*	*Application*	*Remarks*
Low-carbon steel (mild steel)	Good tensile strength, ductility and toughness. Reasonably malleable. High elasticity	Available in many forms and sections, e.g. black, bright, rolled, drawn, squares, rounds, tubes. Used for most general engineering work, as a constructional steel and for the manufacture of bolts, nuts, structural steels (channels, angles, joists)	Not suitable for applications requiring high strength and hardness. Can only be case hardened
Medium-carbon steel	Ductile at the lower end of the carbon content but mainly stronger and harder than low-carbon steel. Not as tough or malleable	Parts requiring greater strength: axles, spindles, crankshaft forgings, cold chisels, leaf springs	Can be hardened to improve wear resistance and toughness

Type	*Property*	*Application*	*Remarks*
High-carbon steel	Combines hardness with high strength. Poor ductility, toughness and malleability	Particularly suitable for cutting tools, drills, taps, dies, files, etc., and coil springs	Can be hardened to improve wear resistance and toughness. Supplied in ground bar form (known as silver steel) and used in the manufacture of high-quality machined parts
Alloy steels	Improvement in various properties dependent upon the element added: e.g. nickel increases tensile strength, toughness, hardness and resistance to fatigue; chromium increases hardness and resistance to corrosion	High-speed steels: used for good-quality cutting tools, e.g. drills, reamers, lathe tools and hacksaws. Stainless steel: used for highly stressed machine parts and heat-resisting applications	Tungsten, chromium and vanadium added Nickel and chromium added
Cast iron	High strength and rigidity in compression. Hard but brittle. Absorbs vibrations. Self-lubricating (from graphite flakes)	Casts readily. Used for machine tool beds and slides, gears and housings	Fairly cheap. Easily machined to a good finish

Non-ferrous metals

Type	*Property*	*Application*	*Remarks*
Copper	Good tensile strength and ductility. Corrosion resistant. High electrical and thermal conductivity	Used mainly for applications where its conductivity property is most useful, e.g. electric wire and cable, boiler tubes, etc.	
Brass (copper-based alloy)	High resistance to corrosion. The higher copper content brasses are quite ductile and suitable for drawing and pressing. The lower copper content brasses are harder and stronger but less ductile and are suitable for casting and forging	Obtainable in rods, bars, sheets and tubes. Used for the manufacture of cartridge cases, small gears and car radiator shells. Also for the production of castings, forgings and pressings	Alloy of copper and zinc. Machines easily

Type	Property	Application	Remarks
Bronze (copper-based alloy)	Hardwearing. Resistant to corrosion. Harder and tougher than brass	Obtainable in rods, bars, sheets and tubes. Used for bearings, springs, piping	Alloy of copper and tin. Easily machined. When phosphorus is added it is known as phosphor bronze which is used predominantly for bearings and gears
Aluminium alloys	In its pure state aluminium is quite soft and has low strength. More useful as an engineering material when alloyed with other elements such as copper, nickel or chromium to increase strength and hardness. Resistant to corrosion; ductile and malleable. High electrical and thermal conductivity	Some aluminium alloys are readily cast and can be used where a corrosion-resistant, light, low-stressed material is required, e.g. gear boxes, crank cases, etc. Also available as rolled bars, sections, sheet and strip. Used for electrical cables and cooking utensils	Duralumin is aluminium alloyed with copper, manganese, magnesium and silicon which gives a high strength-to-weight ratio and is used extensively in the aircraft industry. Aluminium alloys are being used increasingly as a replacement for copper as a conducting material because of their relative cheapness

Non-metallic materials

Type	Property	Application	Remarks
Plastics	Resistant to corrosion. Good insulation properties. Colourful, attractive and pleasant to handle	PVC: bottles, buckets, bowls and electrical cable covering. Nylon: bearings, cams, gears	Plastics have a comparatively low strength
Rubber	Flexible, resilient. Good insulation properties	Seals, bearings, anti-vibration mountings, expansion joints, springs, linings	

FREEHAND SKETCHING

During the course of design a designer will use freehand sketches, perhaps in pictorial or orthographic form. It will be useful for the student to practise a few techniques of freehand sketching.

First of all a pencil should be held with freedom and not too close to the point. Initially start by drawing a straight line. Is it easier to draw left to right horizontally? Or from right to left? (Fig. 6.2a.) Is it harder to draw a vertical straight line upwards or downwards? (Fig. 6.2b.)

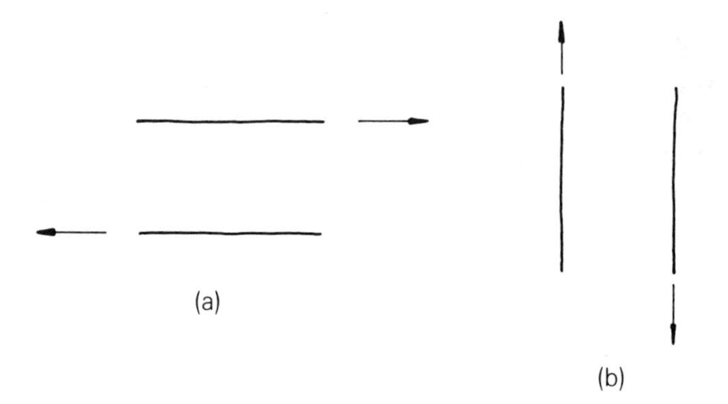

Fig. 6.2 Sketching a straight line

Usually a right-handed person finds it easier to draw horizontally from left to right.

Bearing this in mind, the student may find it easier to construct a rectangle by the method shown in Fig. 6.3.

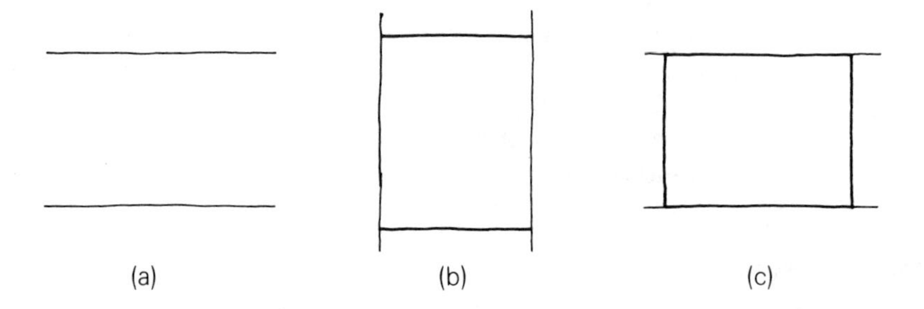

Fig. 6.3 Sketching a rectangle.

(a) Draw two feint horizontal lines at a distance apart equal to the width of the rectangle (Fig. 6.3a).
(b) Turn the paper through 90° and draw two more horizontal lines, at a distance apart equal to the length of the rectangle, to cross the lines previously drawn (Fig. 6.3b).
(c) Turn the paper back through 90°; line-in the feint lines to complete the rectangle (Fig. 6.3c).

The student will know from experience that it is very difficult to make a good free-hand sketch of a circle. With a little pre-planning, however, as shown in Fig. 6.4, the task may not seem so arduous.

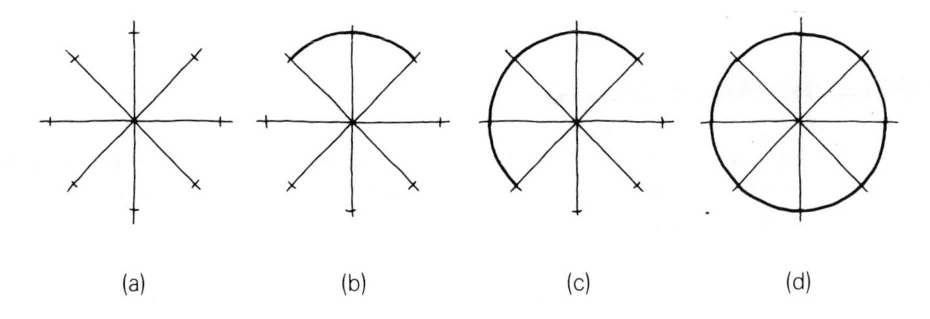

Fig. 6.4 Sketching circles.

(a) Construct the horizontal and vertical centre-lines of the circle (turning the paper through 90° as previously suggested). Also sketch in the intermediate axes, at approximately 45° to the centre-lines, by similar methods. Mark off the radius of the circle on each axis (Fig. 6.4a).
(b) Construct an arc in the position shown (Fig. 6.4b).
(c) Turn the paper through 90° and extend the arc as shown (Fig. 6.4c).
(d) Continue to turn the paper through 90° angles, extending the curve until a complete circle is drawn (Fig. 6.4d).

It will now be useful for the student to practise freehand sketching of simple engineering shapes.

Fig. 6.5a shows a machined block in pictorial view. Using the procedures previously outlined, make freehand sketches of the front and end-views in orthographic projection and in good proportion by estimating all the dimensions. (Fig. 6.5b shows the solution.)

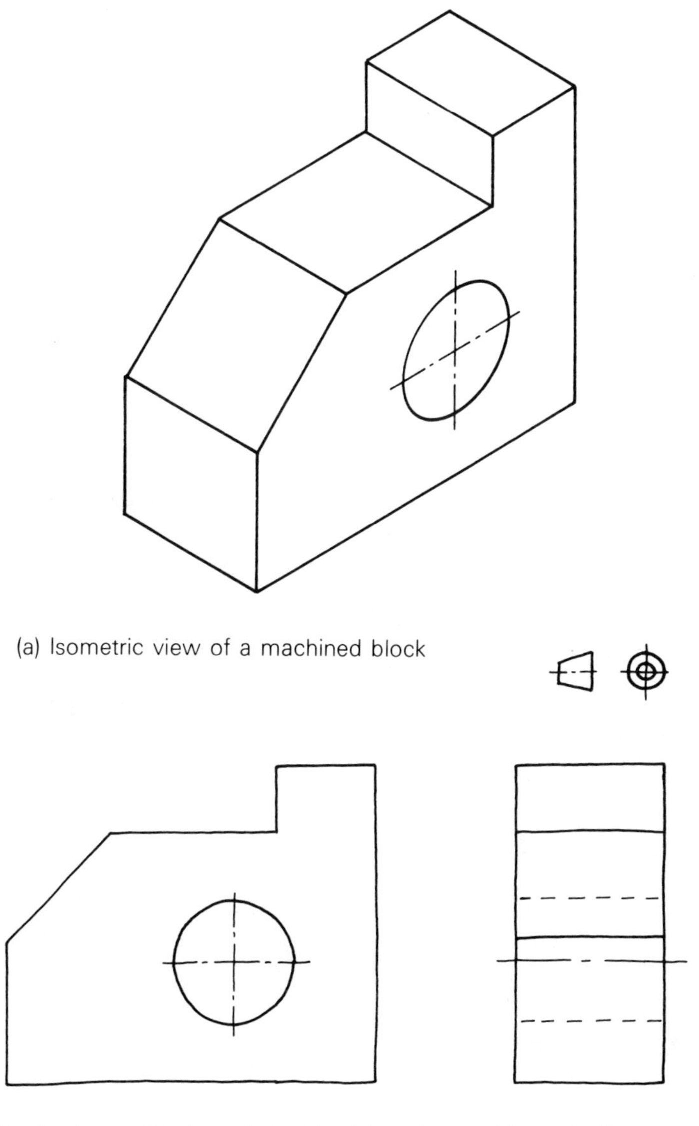

Fig. 6.5 Freehand sketching.

It is often useful to sketch a component pictorially to give a more illustrative view of its form. The student will already be aware of the two systems outlined below.

Sketching in Isometric View

The axes of isometric projection are shown in Fig. 6.6. The lines AB and AC make angles of 30° with the horizontal, AD being perpendicular. A method of constructing

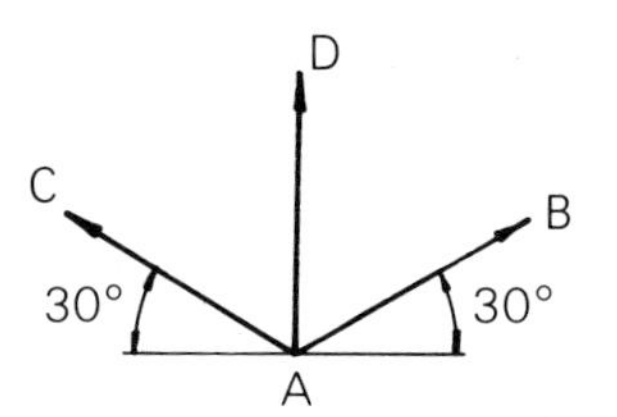

Fig. 6.6 Axes of isometric projection.

circles, in isometric view, is shown in Fig. 6.7. Firstly a plane of projection is established, ABCD, with AB and CD at 30° to the horizontal and AD and BC perpendicular, then:

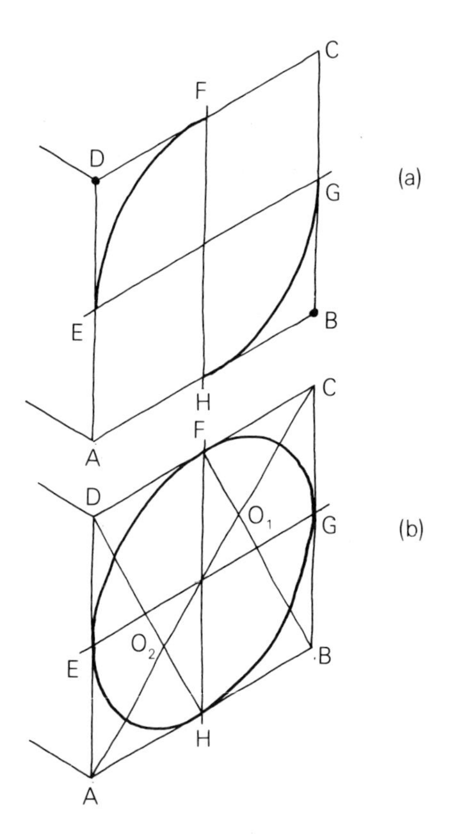

Fig. 6.7 Constructing circles in isometric view.

(a) draw in the centre-lines of the circle, by the method previously outlined (Fig. 6.7a)
(b) with imaginary centre B sketch in the arc EF

(c) turn the paper through 180° and, with imaginary centre D, sketch in the arc GH
(d) draw in the longest diagonal AC (Fig. 6.7b); join the two end-points of the shortest diagonal, B and D, to the mid-points of the opposite sides to produce lines BF and DH, which intersect the diagonal AC at O_1 and O_2 respectively
(e) with imaginary centres O_1 and O_2 draw in the arcs FG and EH to complete the circle in isometric view.

Example 1 Construct an isometric 'crate', approximately 60 mm cube, by freehand sketching. Within each plane construct a circle in isometric view.

The solution is shown in Fig. 6.8.

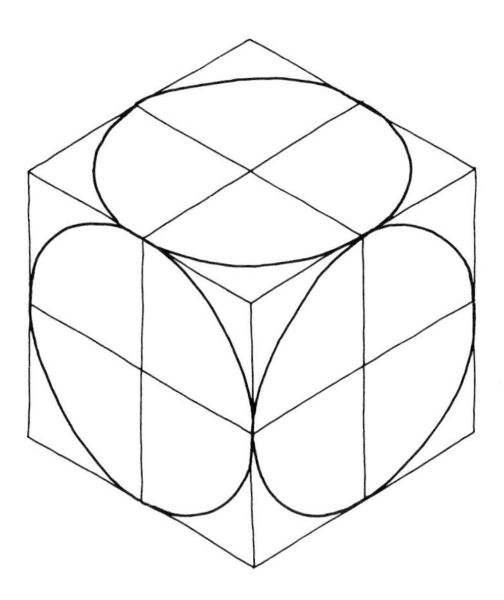

Fig. 6.8 Construction of circles within each isometric plane.

Example 2 Make a freehand sketch, in isometric view, of the angle bracket shown in Fig. 6.9 by the procedure displayed in Figs 6.10a to g, using the methods previously described.

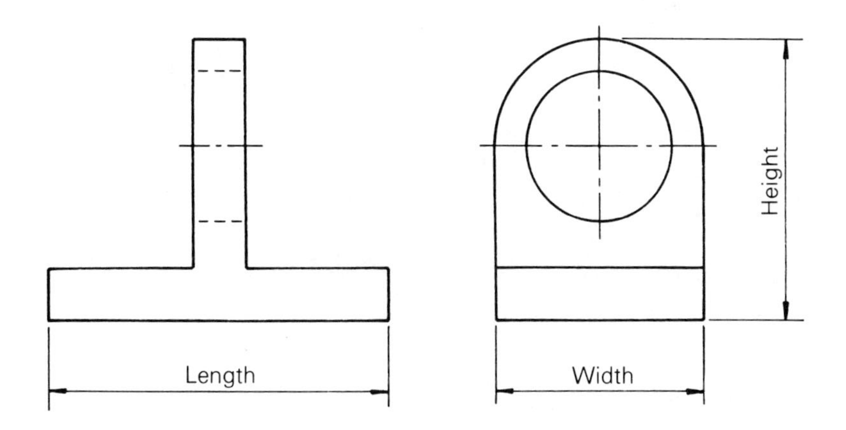

Fig. 6.9 Angle bracket in orthographic projection.

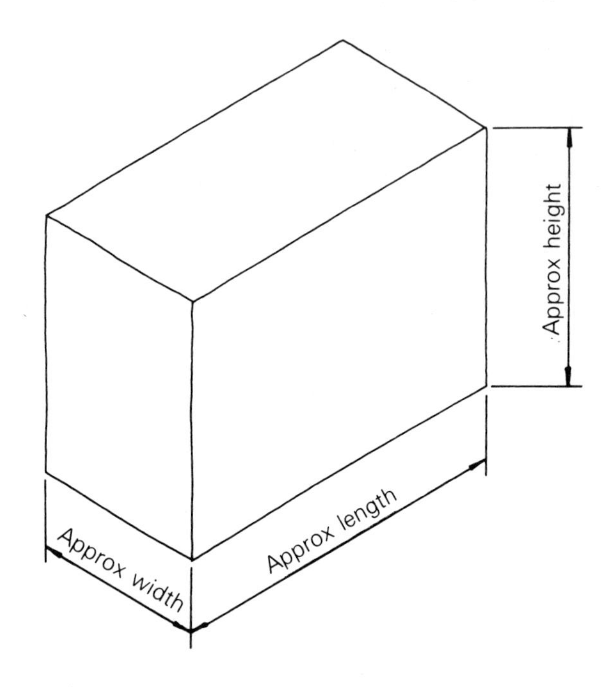

Fig. 6.10(a) Construct an isometric crate

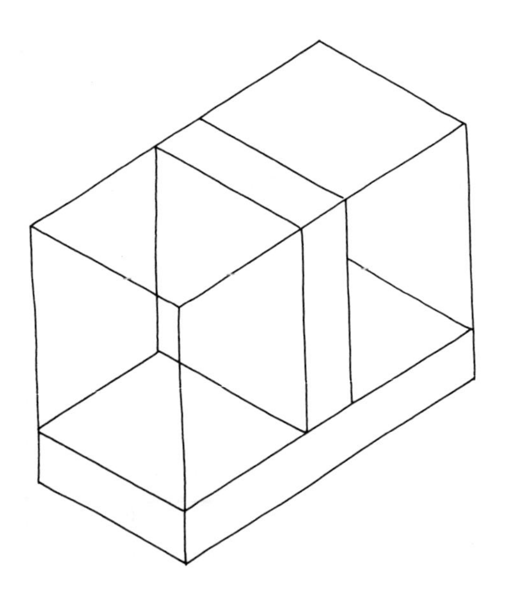

Fig. 6.10(b) Sketch in the basic outline of the angle bracket.

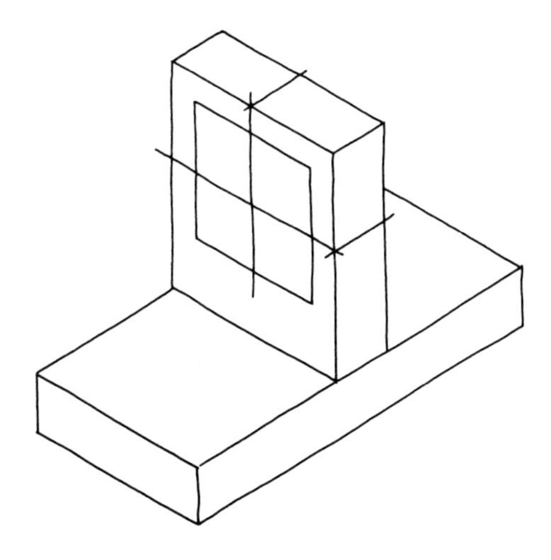

Fig. 6.10(c) Sketch in the outline for the circle and semi-circle.

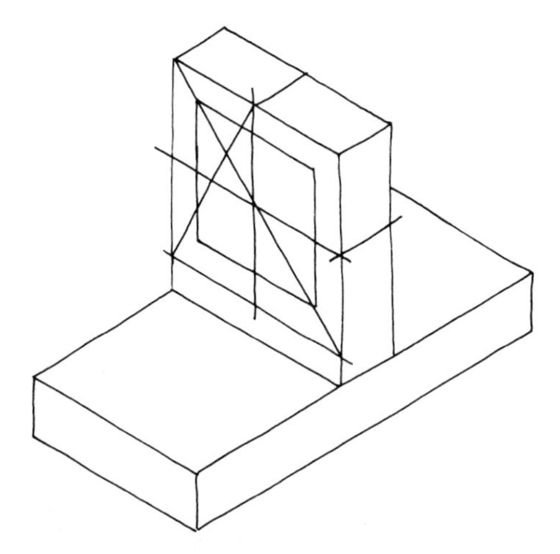

Fig. 6.10(d) Sketch in the construction lines for the semi-circle.

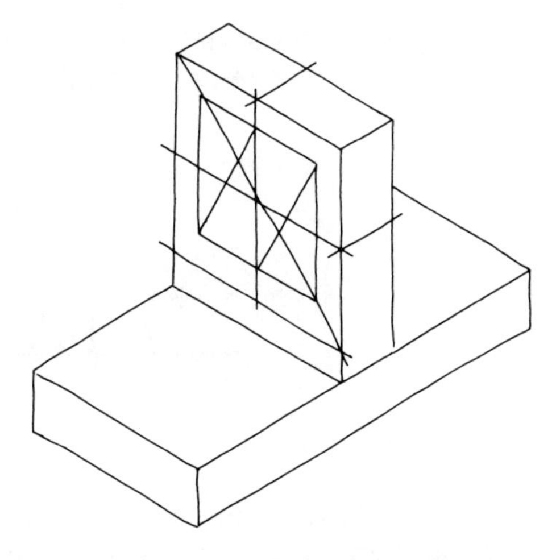

Fig. 6.10(e) Sketch in the construction lines for the circle.

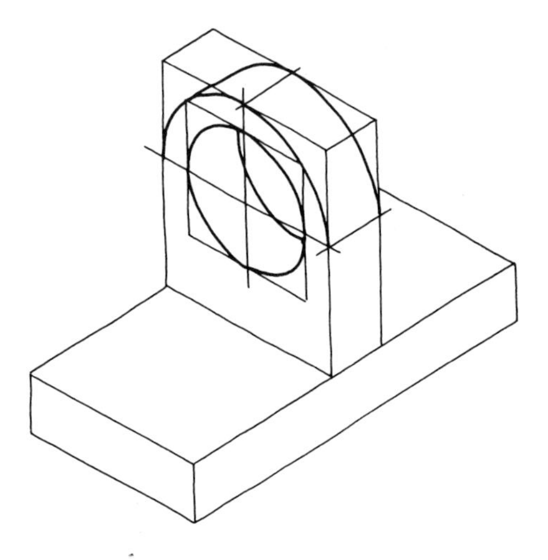

Fig. 6.10(f) Construct the circle and semi-circle.

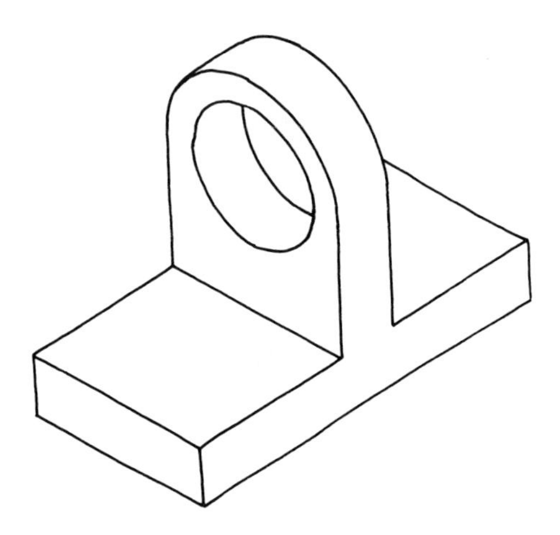

Fig. 6.10(g) Line-in the whole component.

Sketching in Oblique View

The axes of oblique projection are shown in Fig. 6.11. The lines AB and AC are horizontal and perpendicular respectively; AD makes an angle, usually 45°, with the horizontal.

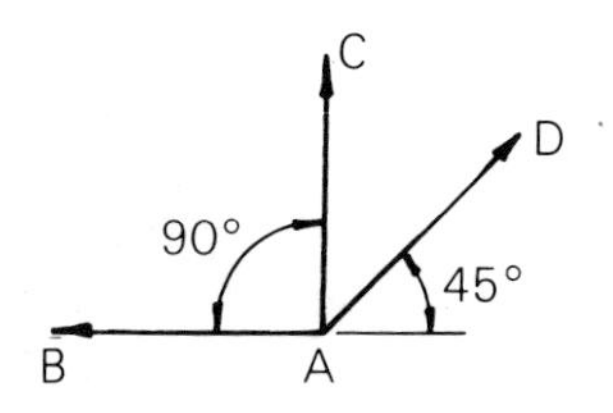

Fig. 6.11 Axes of oblique projection.

In this method of projection one face of the object, and all faces parallel to it, are shown in true shape and size. The lengths in the plane of the sloping side are usually drawn half-size.

Example 3 Make a freehand sketch, in oblique view, of the angle bracket shown in orthographic projection in Fig. 6.9 by the procedure displayed in Figs 6.12a to e using the methods previously described.

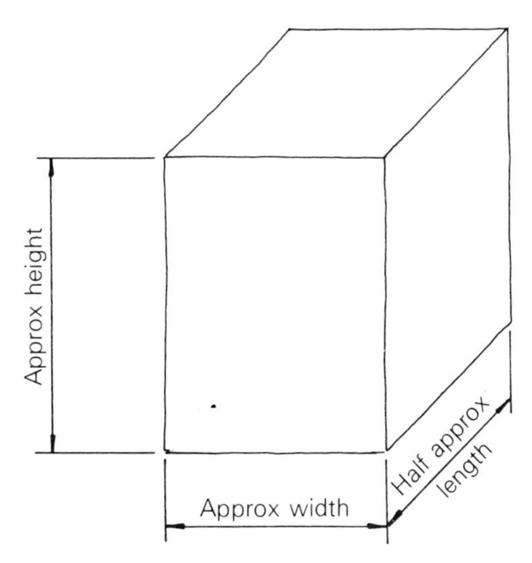

Fig. 6.12(a) Construct an oblique crate.

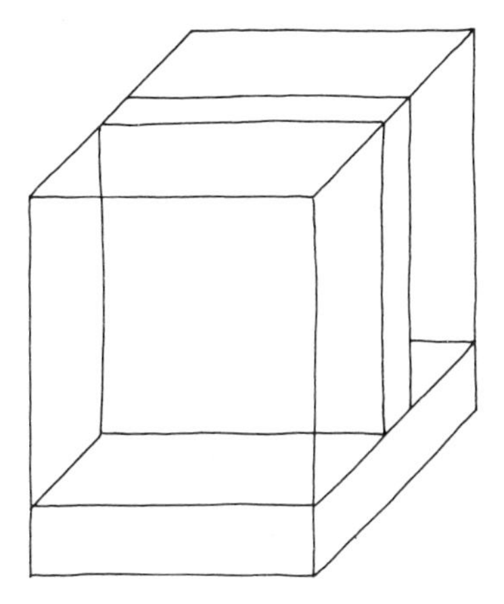

Fig. 6.12(b) Sketch in the basic outline of the angle bracket.

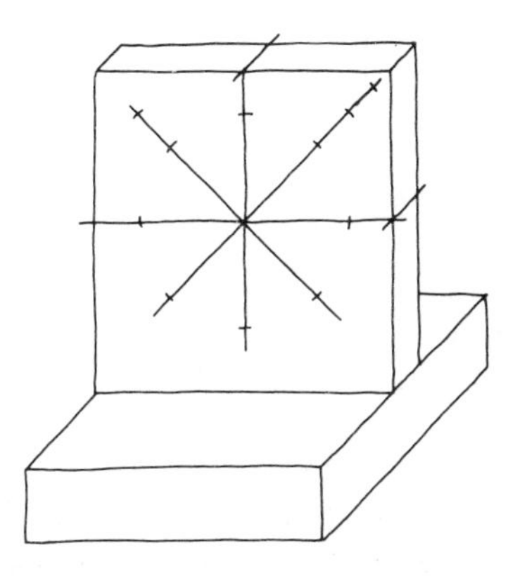

Fig. 6.12(c) Sketch in the construction lines for the circle and semi-circle.

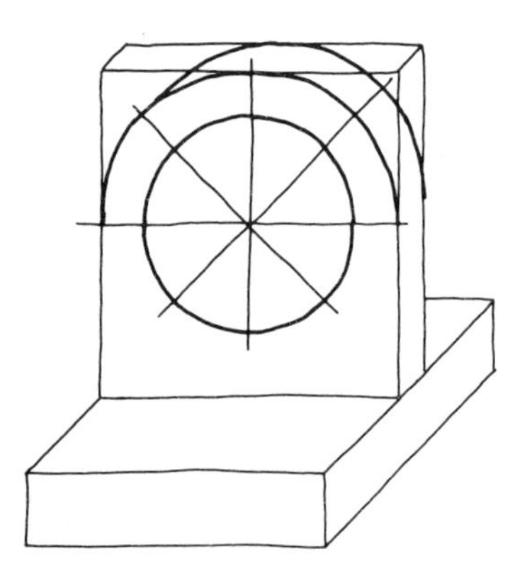

Fig. 6.12(d) Construct the circle and semi-circle.

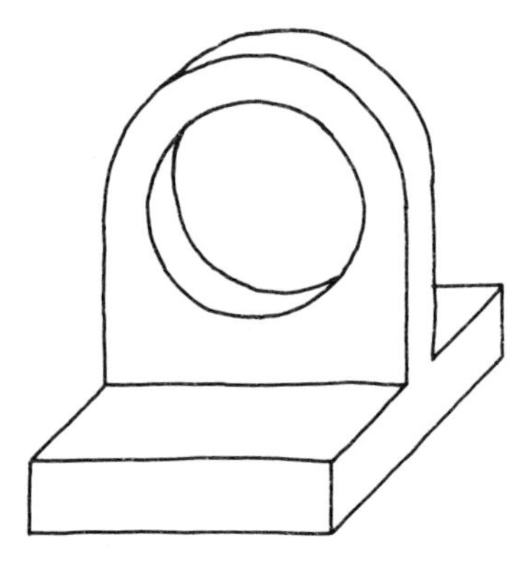

Fig. 6.12(e) Line in the whole component.

PRODUCING ENGINEERING DRAWINGS

An engineering drawing is the means of communicating the ideas of the designer through the draughtsman to all the departments involved in the manufacture of a product. It is essential that the information which it conveys must be in the clearest and most precise form to reduce the possibility of error at any stage of production.

A British Standard Specification, BS 308 Engineering Drawing Practice, was first introduced by the British Standards Institution in 1927 to regularise the approach to the production of engineering drawings. The Specification has been revised from time to time, the latest revision being in 1972 principally to bring the standard in line with published recommendations of the International Organization for Standardization (ISO) on engineering drawing practice, and in particular to introduce, as a standard convention, the system of geometrical tolerance symbols specified by ISO recommendations.

Title Block

The title block is generally preprinted on the original drawing and contains the essential information which is required for the identification, administration and interpretation of the drawing.

The title block should be located at the bottom of the sheet with the drawing number in the bottom right-hand corner. Most large engineering companies have developed their own house style of title block which contains information pertaining to their particular requirements.

BS 308: Part 1: 1972 recommends that spaces should be provided for the following information.

1. Basic information

(a) Name of firm
(b) Drawing number
(c) Title of part or assembly
(d) Original scale
(e) Date of drawing
(f) Signature(s)
(g) Alterations
(h) Copyright clause
(i) Projection symbol (not necessarily in the title block)
(j) Unit of measurement
(k) Reference to drawing practice standards.

2. Additional information

The following list of items of additional information should be considered for inclusion:

(a) Material and specification
(b) Related specifications
(c) Treatment/hardness
(d) Finish
(e) Surface texture
(f) General tolerances
(g) Key to geometrical tolerancing
(h) Screw thread forms
(i) Sheet size
(j) Sheet number
(k) Number of sheets
(l) First used on
(m) Similar to
(n) Equivalent part
(o) Supersedes
(p) Superseded by
(q) Tool references
(r) Gauge references
(s) Grid system or zoning
(t) Warning notes, e.g. 'DO NOT SCALE'
(u) Print folding marks.

Other additional information may also be added.

Fig. 6.13 shows a typical layout of a drawing sheet.

DRG NO.

COPYRIGHT NOTE

©

TITLE

Fig. 6.13 Typical layout of a drawing sheet.

PROJECTION

ISSUE	DATE	CHANGE

DIMENSIONS IN MILLIMETRES (mm) SURFACE TEXTURE VALUES IN μm TOLERANCES UNLESS OTHERWISE STATED DIMENSIONAL ± 0.4 ANGULAR $\pm 0°30'$	MATL SPEC TREATMENT FINISH	DRAWN / CHECKED TRACED / CHECKED APPROVED / DATE
NAME OF COMPANY	SCALE	DRG NO.

Scales

All drawings should be drawn in proportion, i.e. to a uniform scale, and the scale used should be stated on the drawing as a ratio, for example, original scale 1:2. If desired, a warning against scaling the drawing may be given (see Note 2t, p. 135).

Recommended scale ratios

Scale multipliers and divisors of 2, 5 and 10 are recommended. For example:

Multipliers:

1000:1	50:1	20:1 on a drawing means the component/assembly has been drawn at twenty times actual size.
500:1	20:1	
200:1	10:1	
100:1	5:1	
	2:1	

Full size:

1:1	This ratio shows the component/assembly has been drawn to actual size.

Divisors:

1:2	1:50	1:50 on a drawing means the component/assembly has been drawn at one-fiftieth of actual size.
1:5	1:100	
1:10	1:200	
1:20	1:500	
	1:1000	

The selection of a suitable scale for a given component can be a difficult exercise for the beginner. With experience, the skill to ease selection will be developed. The general approach is to decide on a scale which will allow the component to be drawn to a size so that the shape, form and detail can be readily understood. Sufficient space should also be left all round each view of the component to allow dimensions and notes to be added.

Auxiliary Views

Auxiliary views are specially projected views which are constructed to clarify the details of a face or feature.

Fig. 6.14a shows a simple block with an angled face which contains a hole machined normal to it. The block is drawn in orthographic projection in Fig. 6.14b and it is readily apparent that none of the views—front, end, or plan—shows the angled face in true view.

An auxiliary view is constructed in Fig. 6.14c, which is a view normal to the angled face, and shows all details of the face in true view.

Similarly in Fig. 6.15a an isometric view of an angled housing is drawn. In orthographic projection (Fig. 6.15b) the true shape of the square hole is not depicted. It could be shown by the construction of an auxiliary view (Fig. 6.15c).

To construct an auxiliary view:

(a) Construct the necessary views (in this case front and end-views) in orthographic projection.
(b) Choose a reference plane in the end-view (XX). This can be located in any position convenient for measuring the distances to the various points, A, B, C, D, etc.

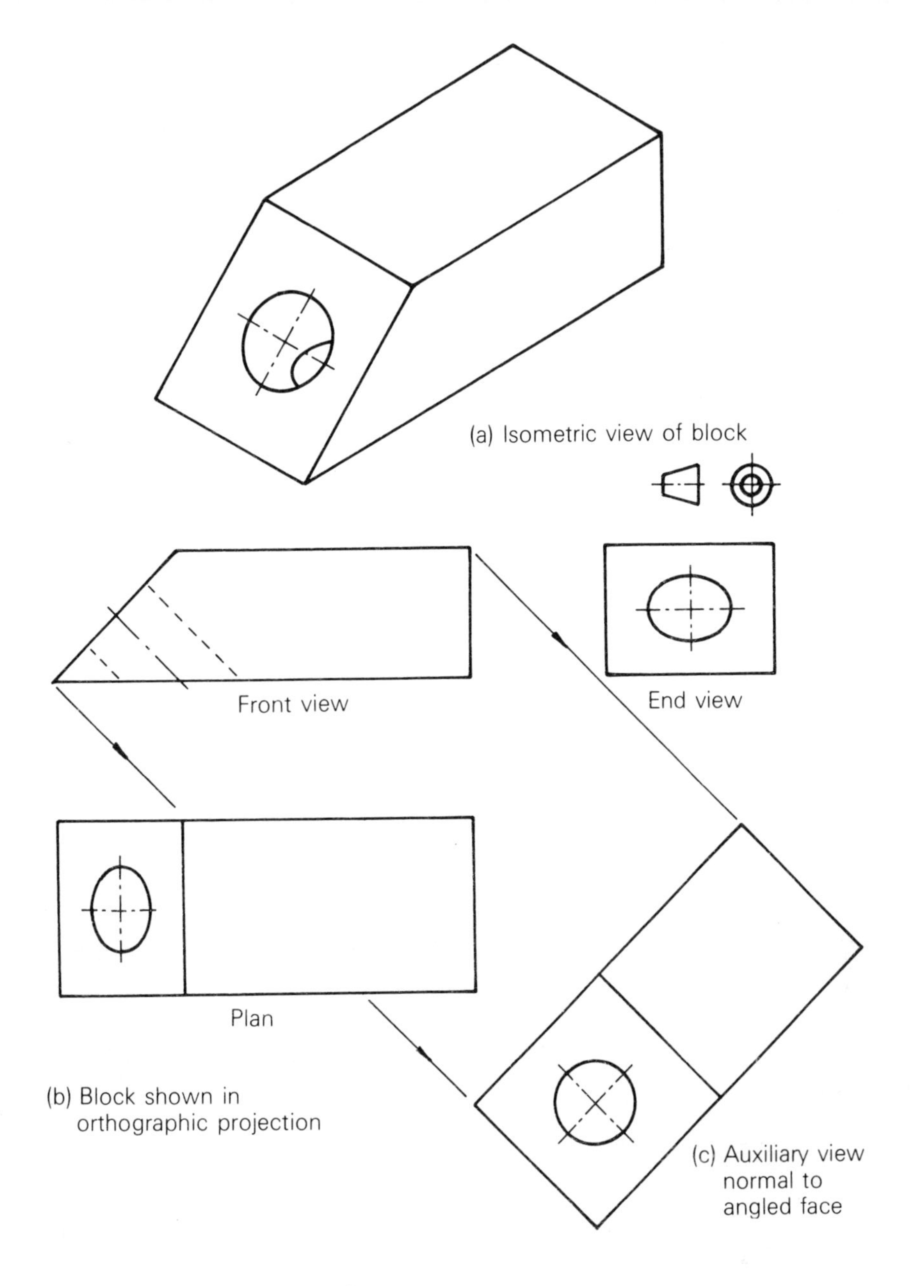

Fig. 6.14 Construction of auxiliary view.

(c) Construct a reference plane (XX) for the auxiliary view, in a convenient position, parallel to the angled face.
(d) Extend construction lines from each feature on the front view to cross the reference plane.
(e) Using dividers or compasses transfer the distances of each point, from the reference plane on the end-view, to the appropriate line on the auxiliary view.
(f) Join up the appropriate points to complete the auxiliary view.

An arrow and a note may be added to show the direction from which the auxiliary view has been constructed. These may be omitted if the meaning is clear without them.

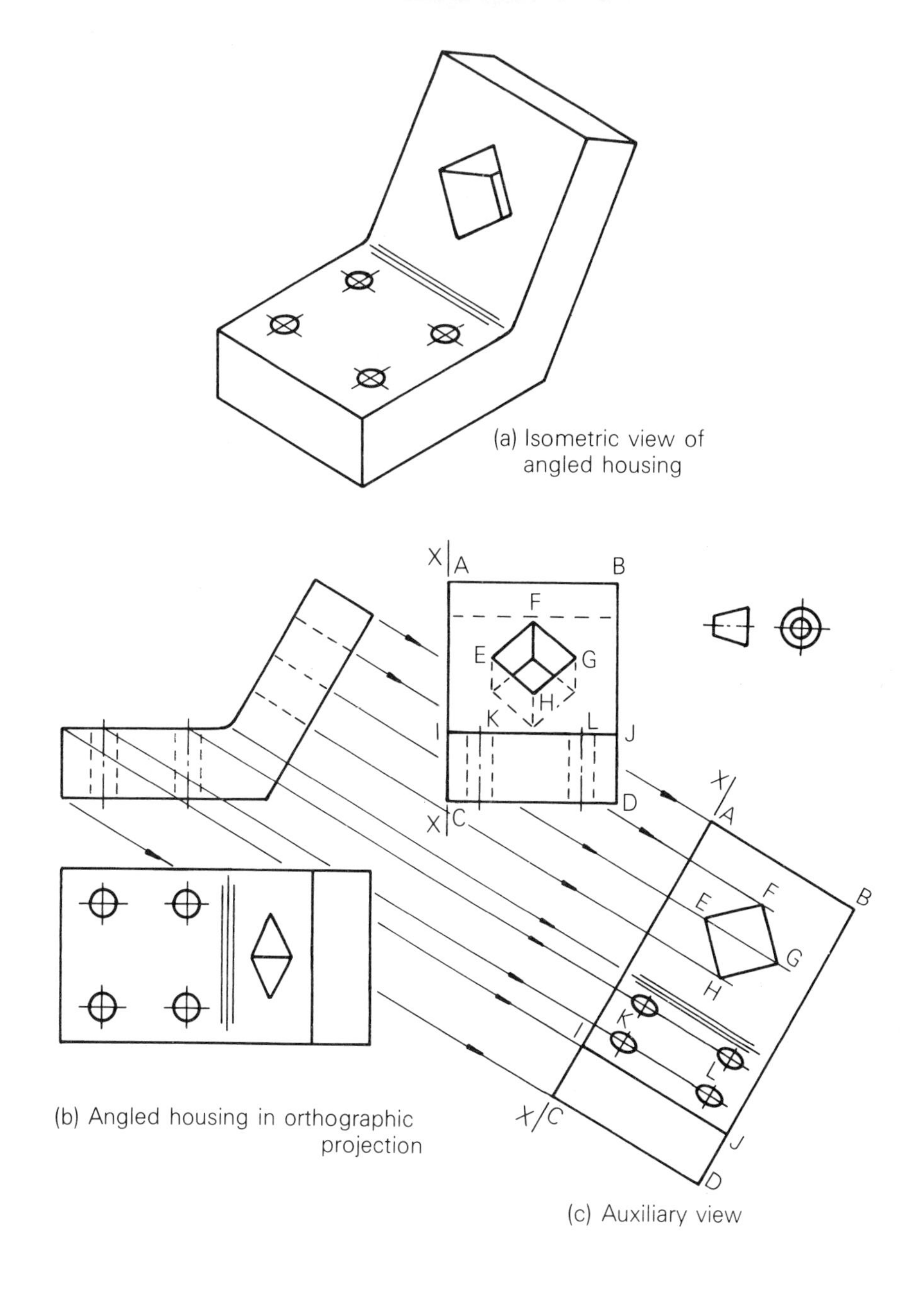

Fig. 6.15 Construction of auxiliary view.

PRODUCING AN ENGINEERING DRAWING – A CASE STUDY

Fig. 6.16 shows a cast iron front cover which is used on a hydraulic cylinder. During the course of manufacture (batch quantities of 100 are produced) four holes of diameter 12.7 mm, are drilled through the casting. A drill jig is required for this operation.

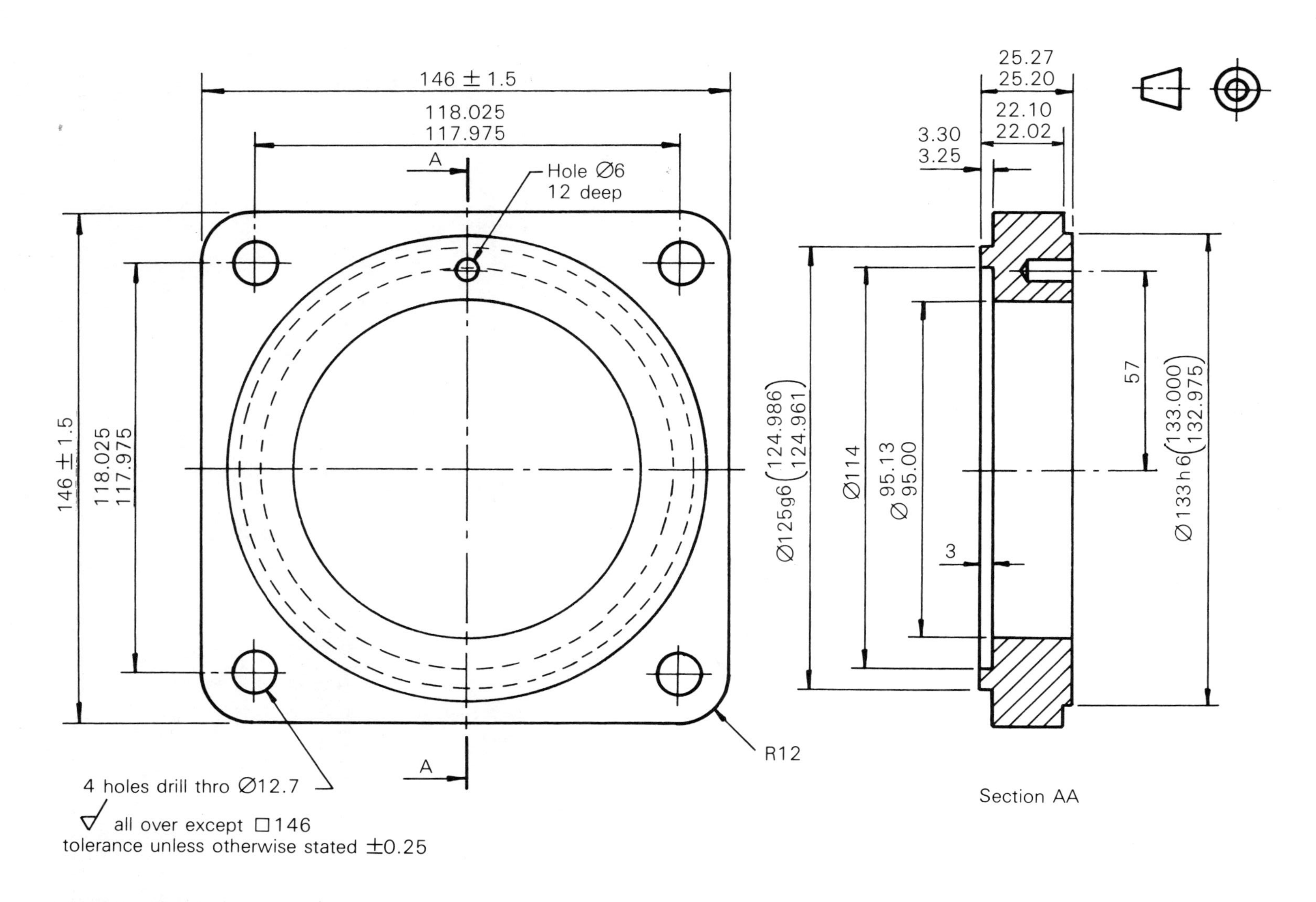

Fig. 6.16 Front cover.

The jig designer may make several freehand sketches before deciding upon the optimum design—see Fig. 6.17.

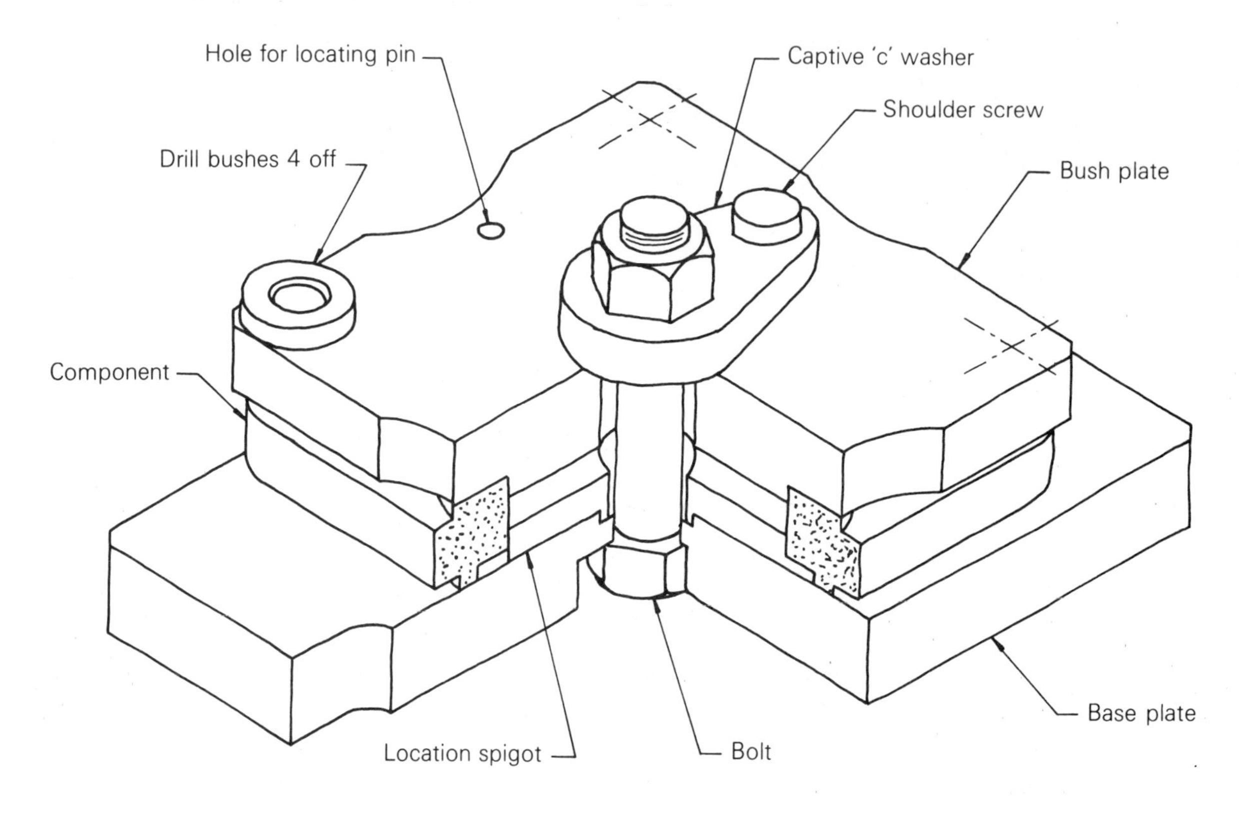

Fig. 6.17 Freehand sketch of proposed drill jig.

From this sketch the detail designer will make an assembly drawing, to scale, of the complete jig with a sufficient number of views (in this case two) to show all the details—see Fig. 6.18.

A *parts list* is added to the drawing to give a description and the quantity of components in the assembly, together with the appropriate drawing number for each part. The student should be aware of the need to use standard parts where possible—this helps to keep costs to a minimum and almost guarantees availability from stock; items 6 to 10 are standard parts. If an assembly drawing consists of many components, it may be necessary to draw up a parts list on a separate drawing sheet.

A detail draughtsman will draw each component, giving all dimensions and tolerances and perhaps adding the type of material from which the component is to be made, any treatment and the type of finish—see Figs 6.19, 6.20 and 6.21.

Materials

Item 1 and	**Base plate:**
Item 2	**Bush plate:** Neither plate is highly stressed, but they need to be tough and have good tensile strength. Mild steel satisfactory. To lessen the risk of wear, which could be sustained during the repetitive loading and unloading of components, the plates should be case hardened.

Item 3 **Positioning pin:** This pin will locate the front cover radially and needs to be tough and wear resistant. Use mild steel, case hardened.

Item 4 **Locking nut:** Used to lock the positioning pin. Needs to be tough. Mild steel satisfactory.

Item 5 **Location spigot:** The spigot locates the front cover and needs to have good strength. It is the component which will be subjected to the most wear. Use medium-carbon steel, hardened and tempered.

Items 6 and 7 **Hexagon headed bolt and nut:** Would be made from mild steel. Satisfactory strength, ductility and toughness.

Item 8 **Drill bushes:** Need to be tough and hard because they have to resist the wear imparted by the drill during entry while drilling each hole. High-carbon steel, hardened and tempered. (See BS 1098: 1967.)

Item 9 **Captive C washer:** Mild steel has sufficient strength and toughness for this component. Because it is constantly being moved over the bush plate it should be case hardened.

Item 10 **Shoulder screw:** Mild steel satisfactory. It has sufficient strength and toughness for this application.

SELF-TEST 6

Select the correct option(s). *Note*: there may be more than one correct answer.

1. Which of the following is not regarded as *basic* information for inclusion in a title block?

(a) drawing number
(b) material
(c) scale
(d) date of drawing.

2. Which of the following scale ratios is not recommended by British Standard Specification?

(a) 5:1
(b) 10:1
(c) 20:1
(d) 25:1.

3. Silver steel is a term applied to a:

(a) low-carbon steel
(b) medium-carbon steel
(c) high-carbon steel
(d) alloy steel.

4. A view which is projected to show the true shape of an inclined face is called an:

(a) isometric view
(b) oblique view
(c) auxiliary view
(d) orthographic view.

(Self-test 6 continues on p. 152.)

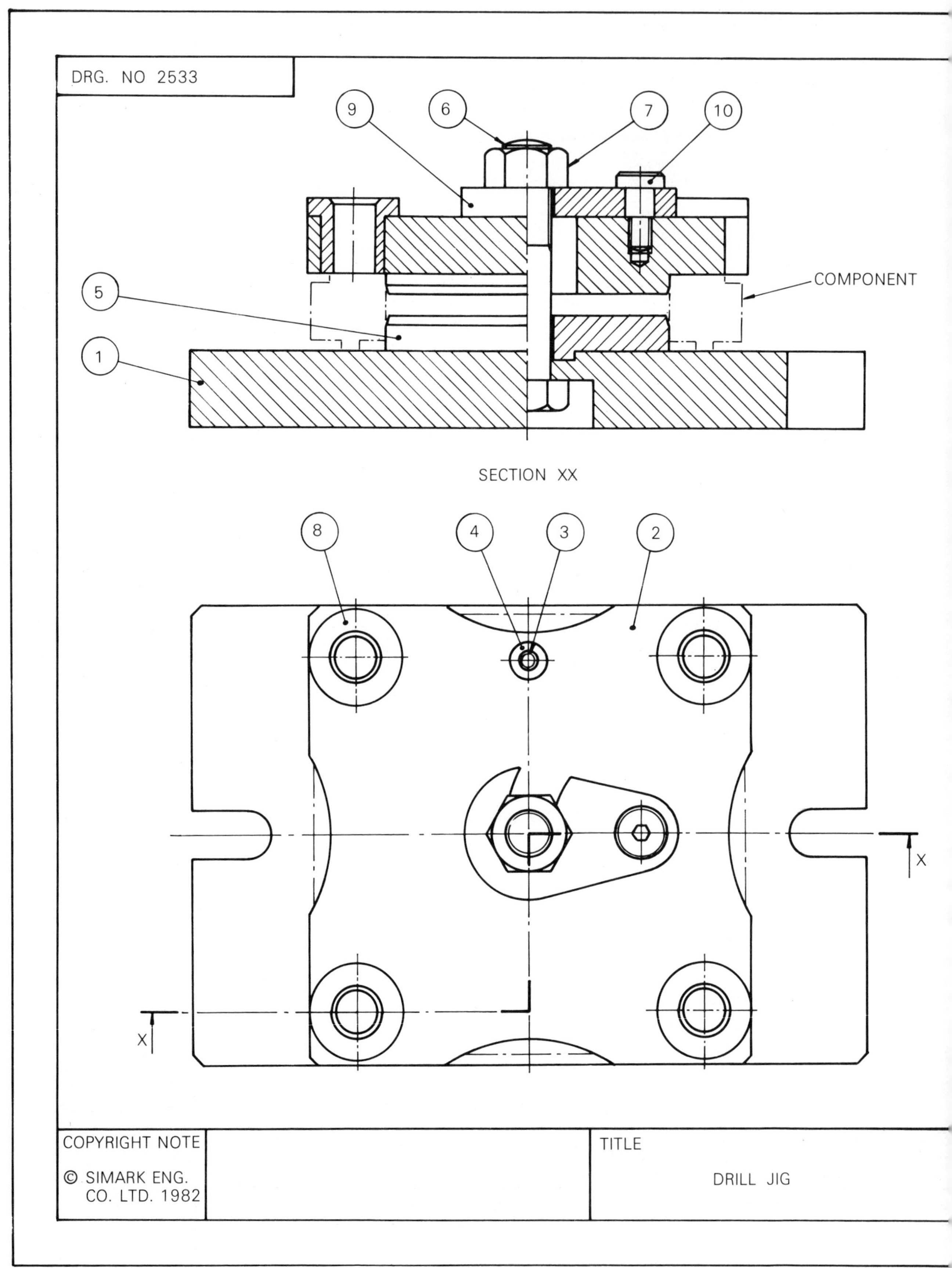

Fig. 6.18 *Assembly drawing for drill jig.*

PROJECTION

ITEM	DRG NO.	DESCRIPTION	QTY
1	2534	BASE PLATE	1
2	2535	BUSH PLATE	1
3	2536	POSITIONING PIN	1
4	2536	LOCKING NUT	1
5	2536	LOCATION SPIGOT	1
6	STANDARD	PRECISION HEX HD BOLT M16 × 2 75 LONG	1
7	STANDARD	PRECISION HEX NUT M16 × 2	1
8	BONEHAM & TURNER STANDARD	SHOULDERED DRILL BUSH SM II. 12.7 × 16	4
9	BONEHAM & TURNER STANDARD	CAPTIVE C WASHER CCW5M	1
10	BONEHAM & TURNER STANDARD	SHOULDER SCREW SS5M	1

ISSUE	DATE	CHANGE

DIMENSIONS IN MILLIMETRES (mm)
SURFACE TEXTURE VALUE IN μm
TOLERANCES UNLESS OTHERWISE STATED
DIMENSIONAL ± 0.4 ANGULAR ± 0°30′

MATL
SPEC
TREATMENT
FINISH

DRAWN		CHECKED SR
TRACED M.S.		CHECKED BJ
APPROVED M.A		DATE 1.6.82

SIMARK ENGINEERING CO LTD. | SCALE 1 : 1 | DRG. NO. 2533

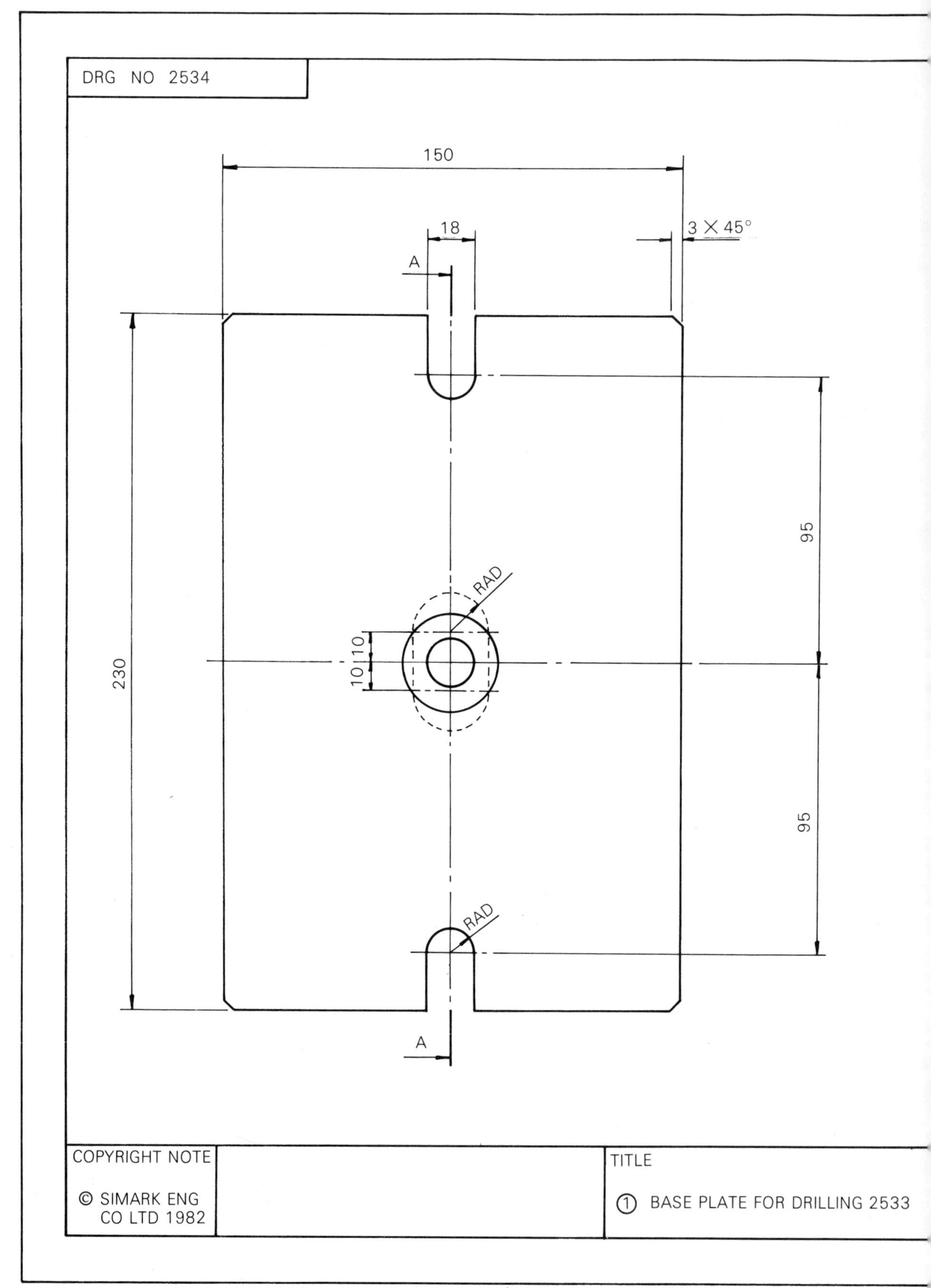

Fig. 6.19 Working drawing of base plate.

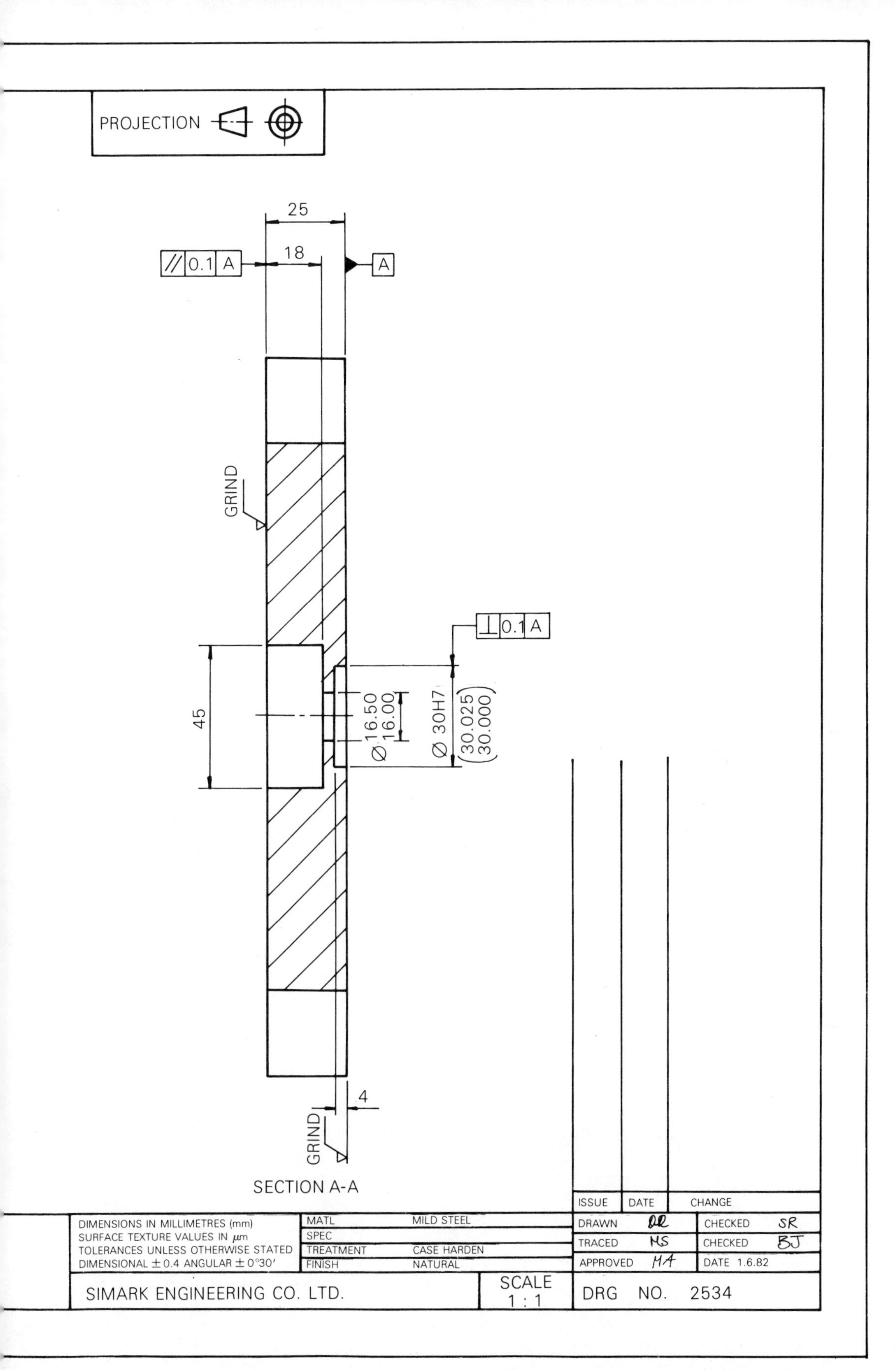

PROJECTION
25
18
0.1 A
A
GRIND
0.1 A
45
Ø 16.50
16.00
Ø 30H7
(30.025
30.000)
4
GRIND
SECTION A-A
ISSUE
DATE
CHANGE
DIMENSIONS IN MILLIMETRES (mm)
SURFACE TEXTURE VALUES IN μm
TOLERANCES UNLESS OTHERWISE STATED
DIMENSIONAL ± 0.4 ANGULAR ± 0°30′
MATL
MILD STEEL
SPEC
TREATMENT
CASE HARDEN
FINISH
NATURAL
DRAWN
CHECKED SR
TRACED MS
CHECKED BJ
APPROVED MA
DATE 1.6.82
SIMARK ENGINEERING CO. LTD.
SCALE
1 : 1
DRG NO. 2534

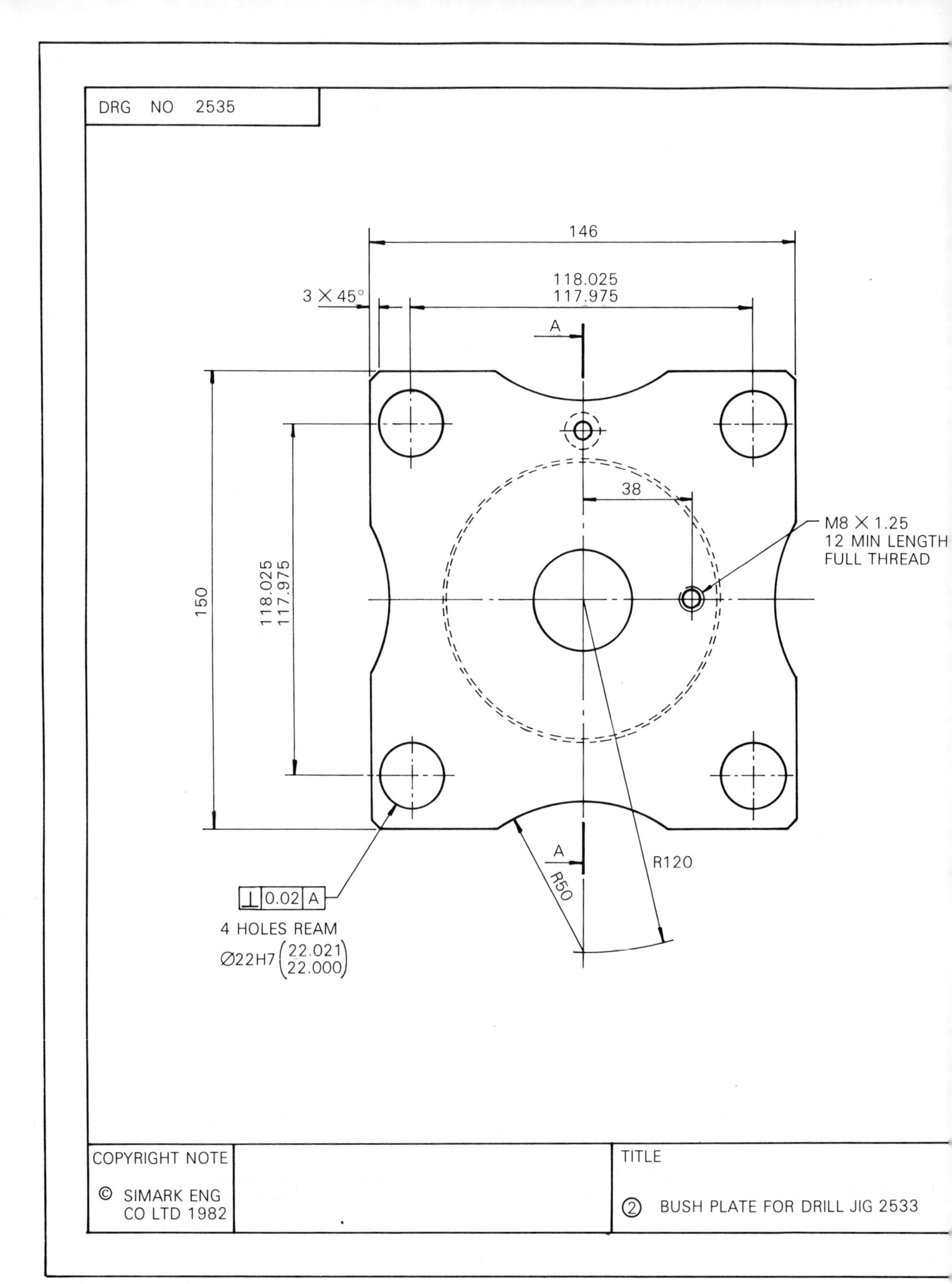

 Fig. 6.20 Working drawing for bush plate.

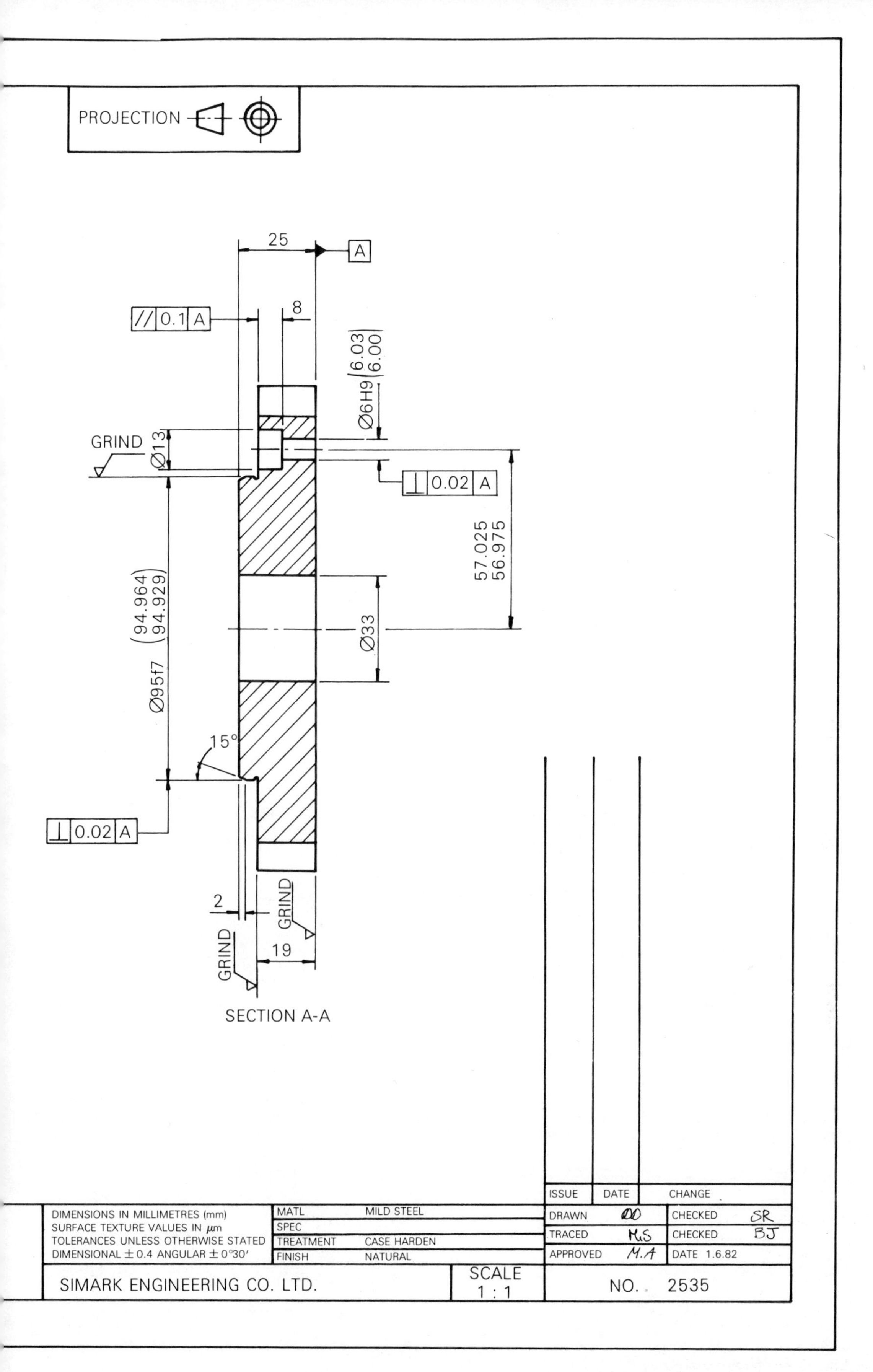
PROJECTION
25
A
// 0.1 A
8
Ø6H9 (6.03 / 6.00)
GRIND
Ø13
⊥ 0.02 A
57.025
56.975
(94.964 / 94.929)
Ø95f7
Ø33
15°
⊥ 0.02 A
2
GRIND
GRIND
19
SECTION A-A
ISSUE
DATE
CHANGE
DIMENSIONS IN MILLIMETRES (mm)
SURFACE TEXTURE VALUES IN μm
TOLERANCES UNLESS OTHERWISE STATED
DIMENSIONAL ± 0.4 ANGULAR ± 0°30′
MATL
MILD STEEL
SPEC
TREATMENT
CASE HARDEN
FINISH
NATURAL
DRAWN
CHECKED SR
TRACED M.S
CHECKED BJ
APPROVED M.A
DATE 1.6.82
SIMARK ENGINEERING CO. LTD.
SCALE
1 : 1
NO. 2535

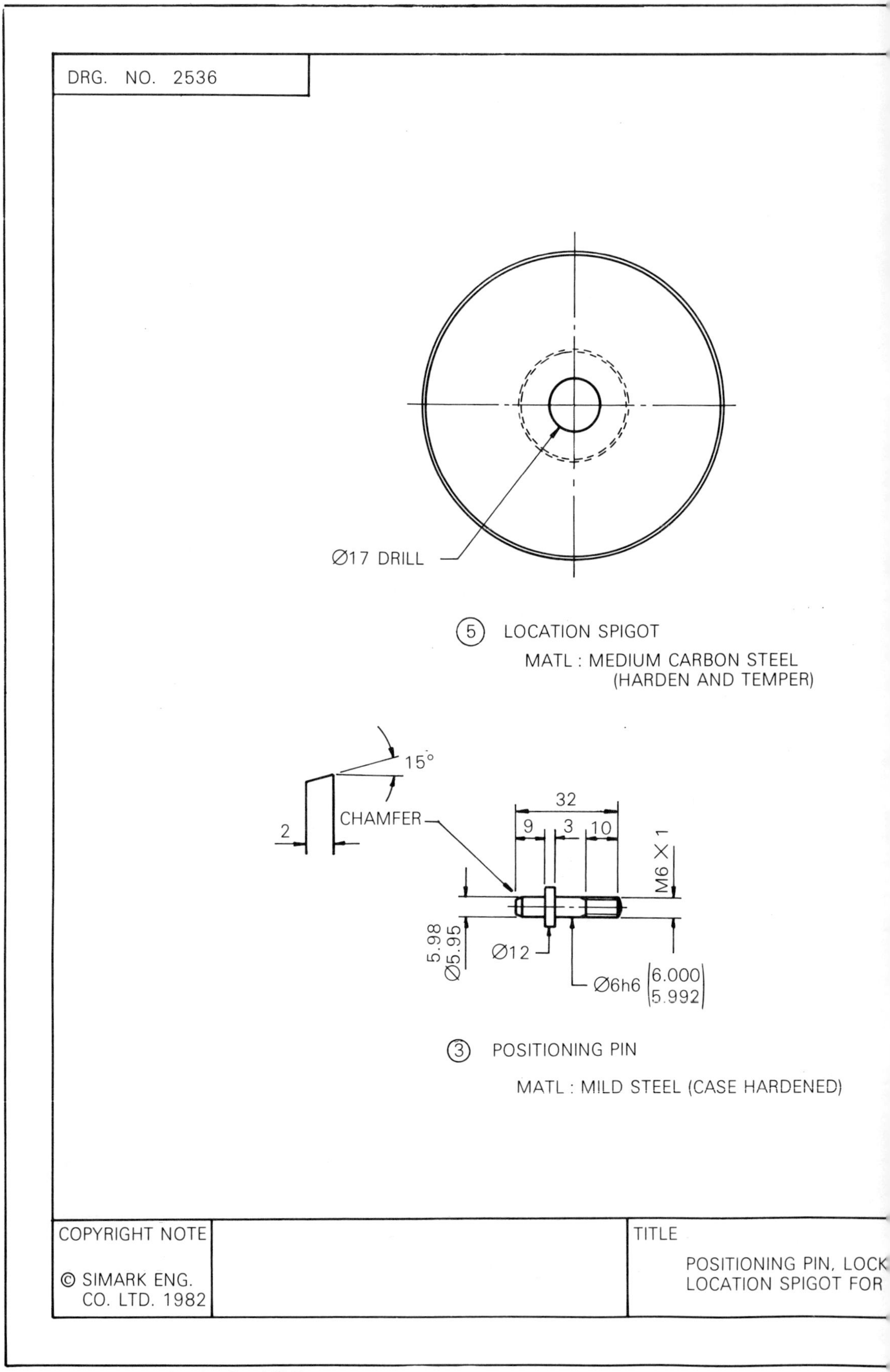

DRG. NO. 2536

Ø17 DRILL

(5) LOCATION SPIGOT

MATL : MEDIUM CARBON STEEL
(HARDEN AND TEMPER)

15°

2

CHAMFER

32

9 3 10

M6 × 1

5.98
Ø5.95

Ø12

Ø6h6 (6.000 / 5.992)

(3) POSITIONING PIN

MATL : MILD STEEL (CASE HARDENED)

COPYRIGHT NOTE		TITLE
© SIMARK ENG. CO. LTD. 1982		POSITIONING PIN, LOCK LOCATION SPIGOT FOR

Fig. 6.21 Working drawing for drill jig components.

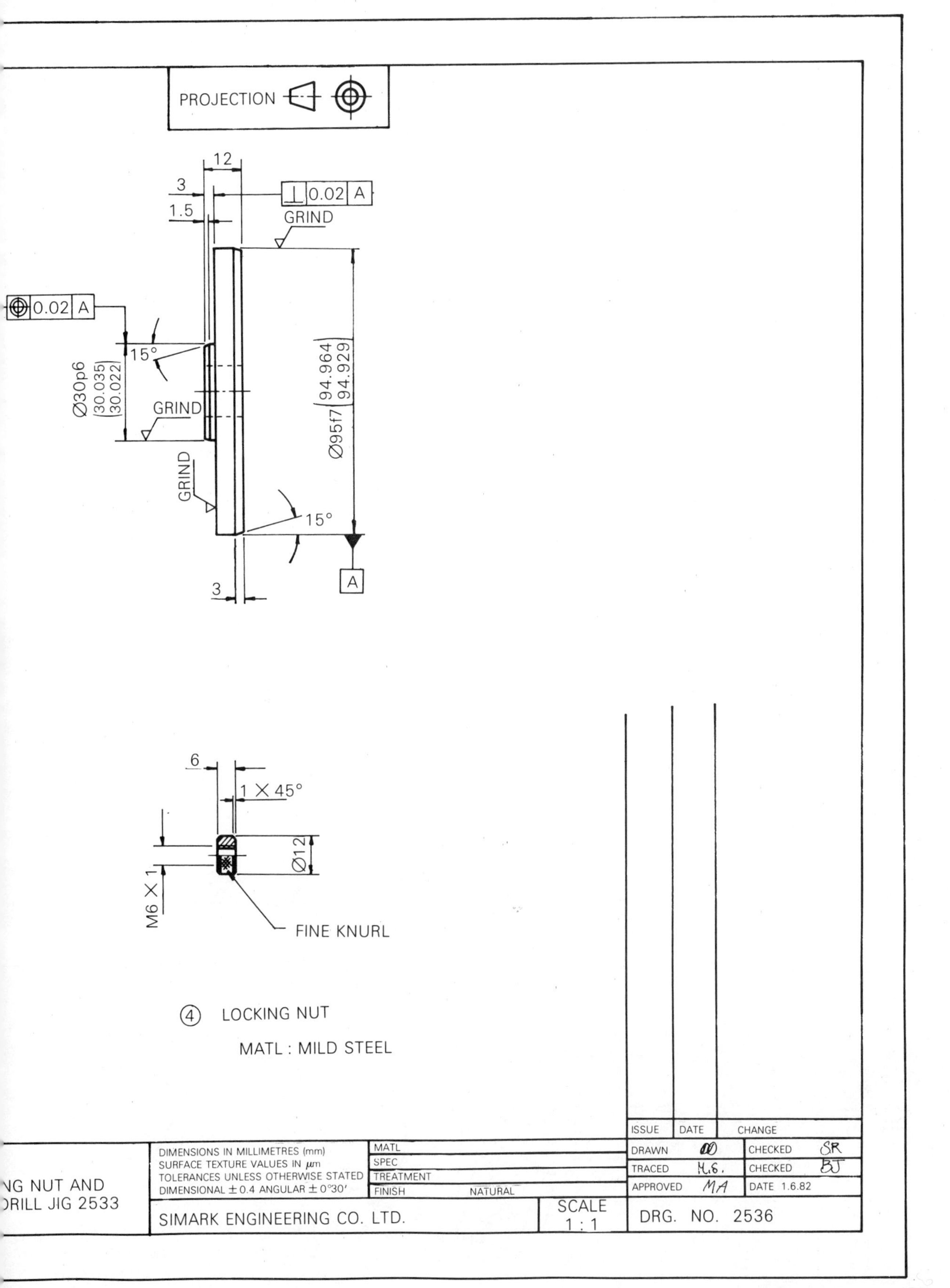
PROJECTION
12
3
1.5
0.02 A
GRIND
0.02 A
15°
Ø30p6
(30.035
(30.022
GRIND
GRIND
Ø95f7 (94.964
94.929)
15°
A
3
6
1 X 45°
Ø12
M6 X 1
FINE KNURL
④ LOCKING NUT
MATL : MILD STEEL
ISSUE
DATE
CHANGE
DIMENSIONS IN MILLIMETRES (mm)
SURFACE TEXTURE VALUES IN μm
TOLERANCES UNLESS OTHERWISE STATED
DIMENSIONAL ± 0.4 ANGULAR ± 0°30′
MATL
SPEC
TREATMENT
FINISH NATURAL
DRAWN
CHECKED SR
TRACED M.S.
CHECKED BJ
APPROVED MA
DATE 1.6.82
NG NUT AND
DRILL JIG 2533
SIMARK ENGINEERING CO. LTD.
SCALE
1 : 1
DRG. NO. 2536

5. The ability of a material to be compressed without fracture means it has the property of:

(a) malleability
(b) ductility
(c) toughness
(d) hardness.

6. One material which lacks toughness is:

(a) cast iron
(b) aluminium alloy
(c) low-carbon steel
(d) alloy steel.

7. The material with the highest tensile strength/weight ratio is:

(a) low-carbon steel
(b) copper alloy
(c) cast iron
(d) aluminium alloy.

8. The correct BS symbol to indicate the system of projection used in Fig. 6.22 is:

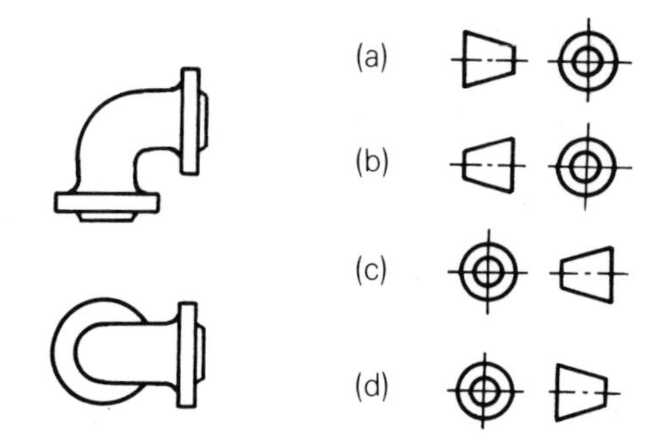

Fig. 6.22

9. Which of the views shown below is incorrect?

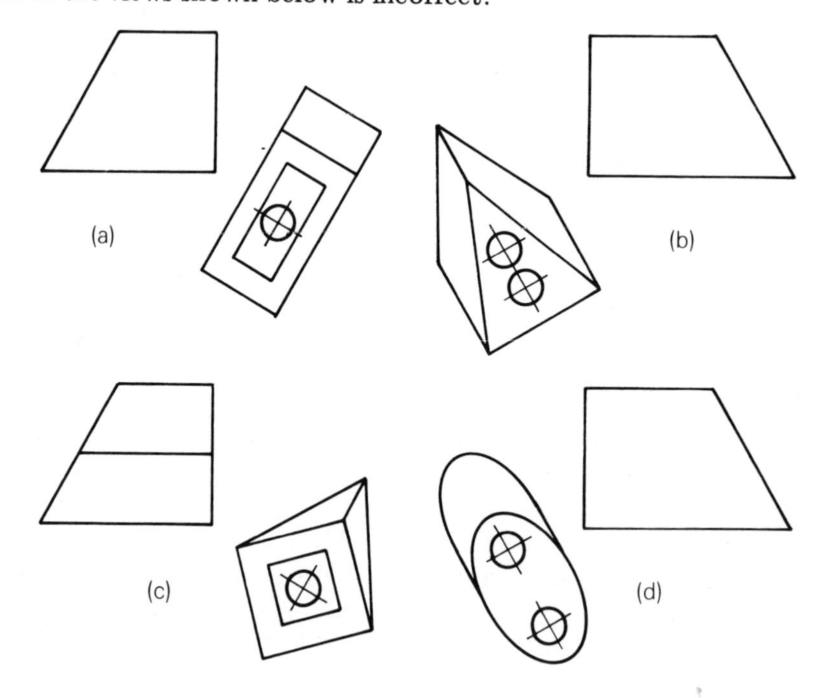

10. The correct BS 308 abbreviation for 'Drawing' is:

(a) DRG
(b) DWG
(c) Drg
(d) Dwg.

EXERCISE 6

1. On A2 drawing paper draw the front view and end-view of the square flange fitting shown in Fig. 6.23 to the sizes and positions stated. Using the reference plane XX, construct an auxiliary view normal to the 45° angled flange. Holes should only be shown in this flange.

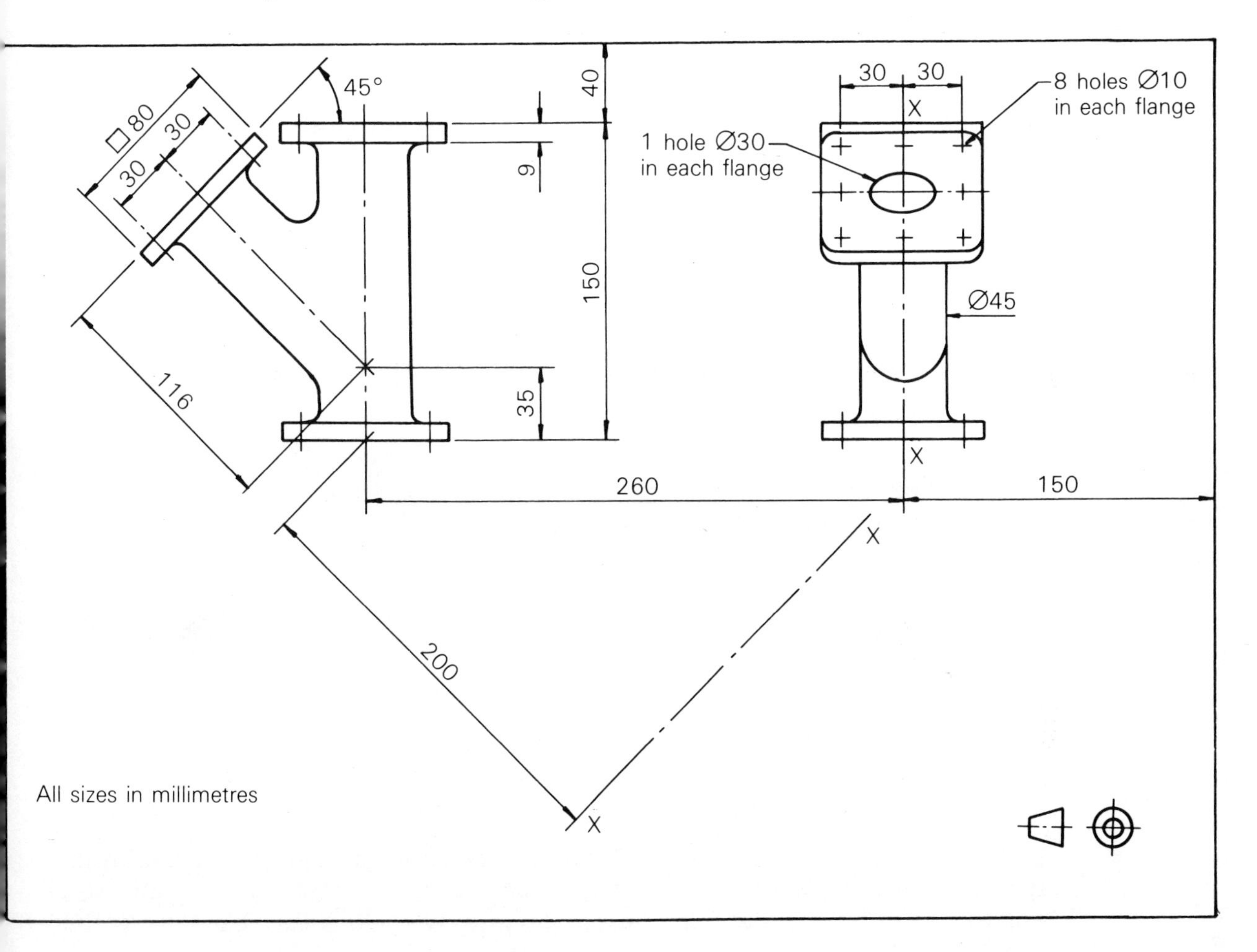

Fig. 6.23 Square flange fitting.

2. On A2 drawing paper draw the front view and end-view of the hexagonal block shown in Fig. 6.24 to the sizes and positions stated. Using the reference plane XX, construct an auxiliary view normal to the 30° angled face.

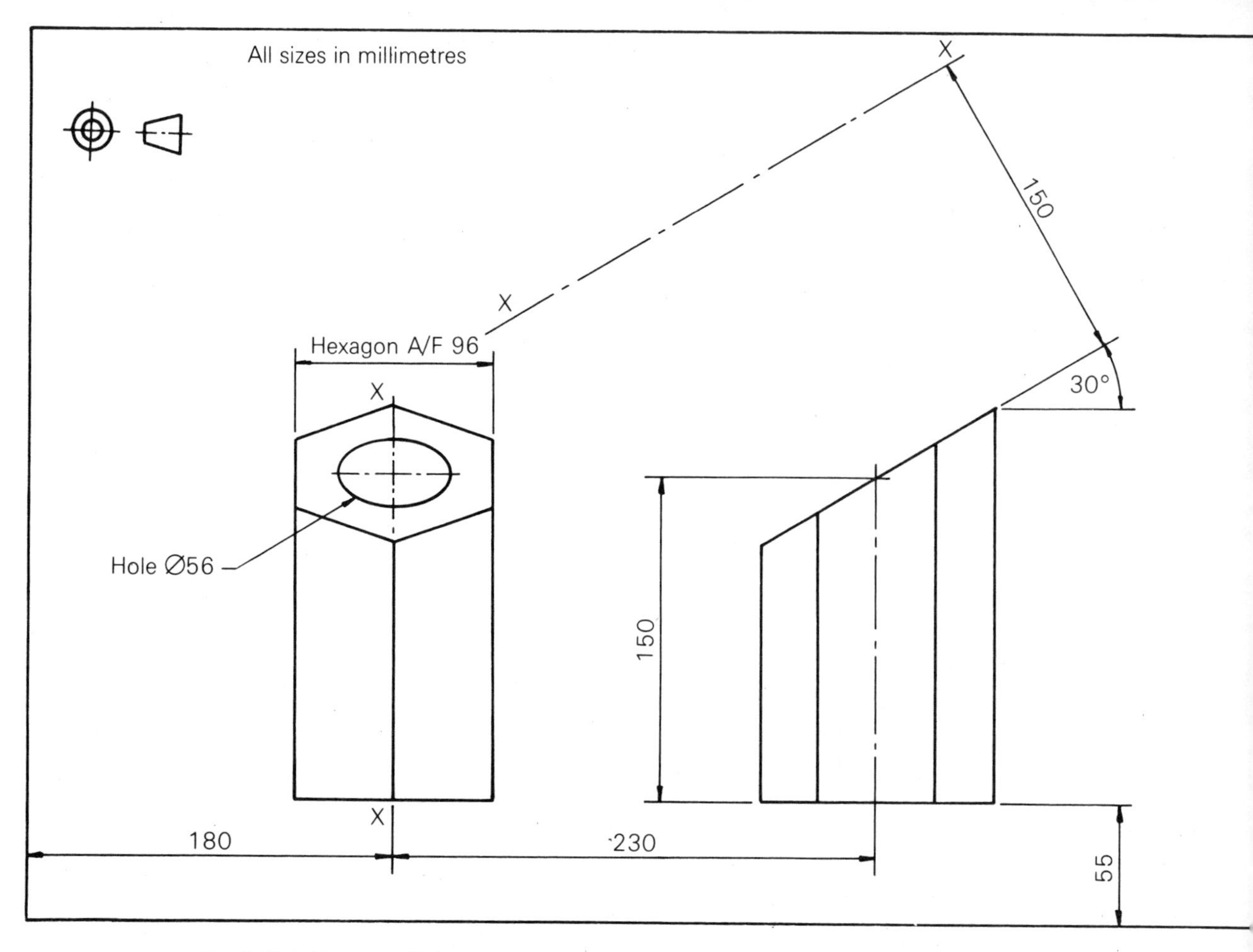

Fig. 6.24 Hexagonal block.

3. A pictorial view of an angle tee-clip is shown in Fig. 6.25. Draw a front view and end-view of the clip in first-angle orthographic projection. Construct an auxiliary view normal to the angled face.

4. A pictorial view of a bracket is shown in Fig. 6.26. Draw three views of the bracket in third-angle orthographic projection. Construct an auxiliary view normal to the square hole.

5. A pictorial view of a drill press vice, together with the component parts, is shown in freehand sketch form in Fig. 6.27. Sketch in good proportion:

(a) the rear end-view of the body in orthographic projection
(b) a pictorial view of the slide looking at the jaw face.

6. For the drill press vice shown in Fig. 6.27, produce a plan and half-sectional front elevation (taken through the centre-line of the main screw) of the vice assembly to a suitable scale. Write a parts list on this drawing and add a title block to British Standard Specification.

7. Produce a fully dimensioned working drawing for each component of the drill press vice in Question 6 to British Standard Specification. Tolerances to be included to BS 308: Part 3: 1972 (ISO). Select suitable material for each component and state which mechanical property of the material makes it appropriate for that use. Add a title block.

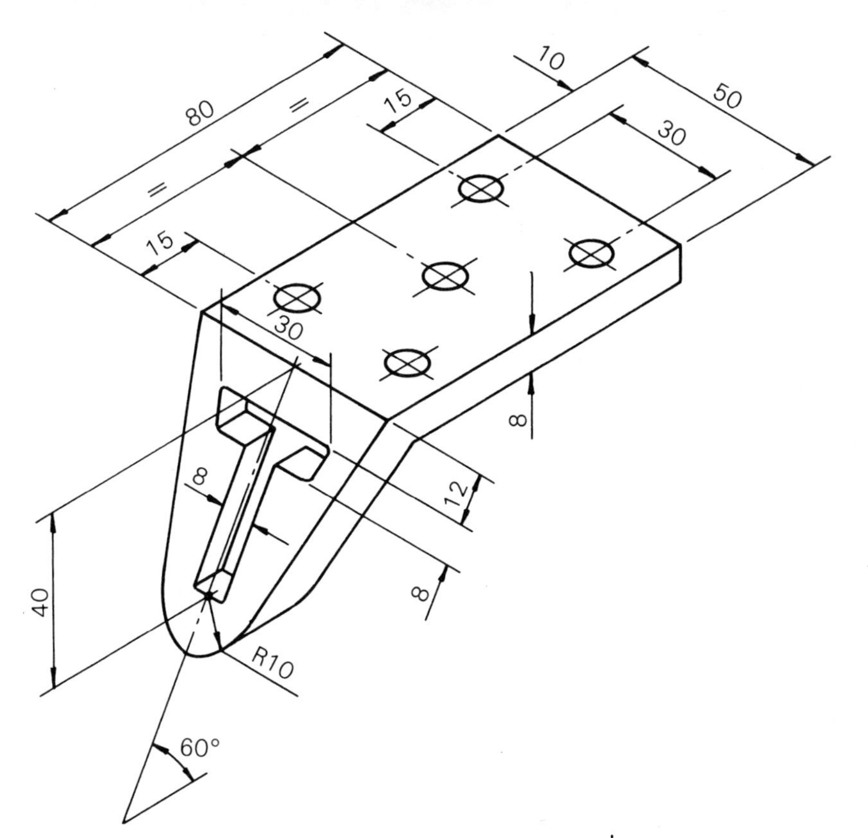

Fig. 6.25 Angle tee-clip.

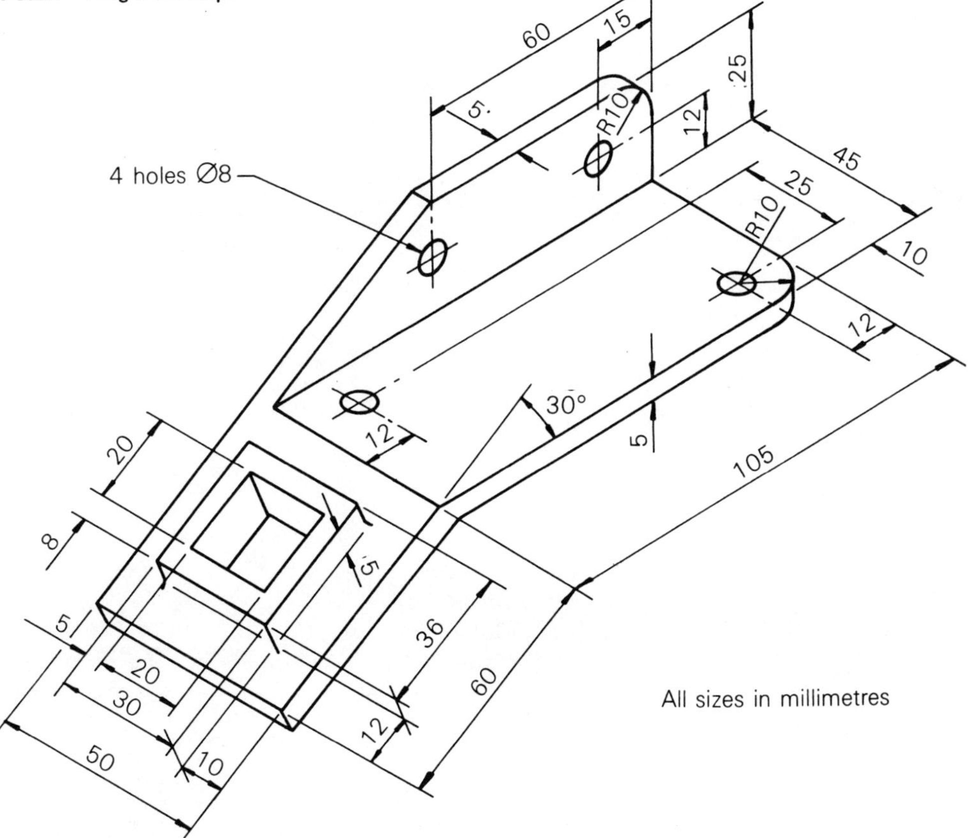

Fig. 6.26 Bracket.

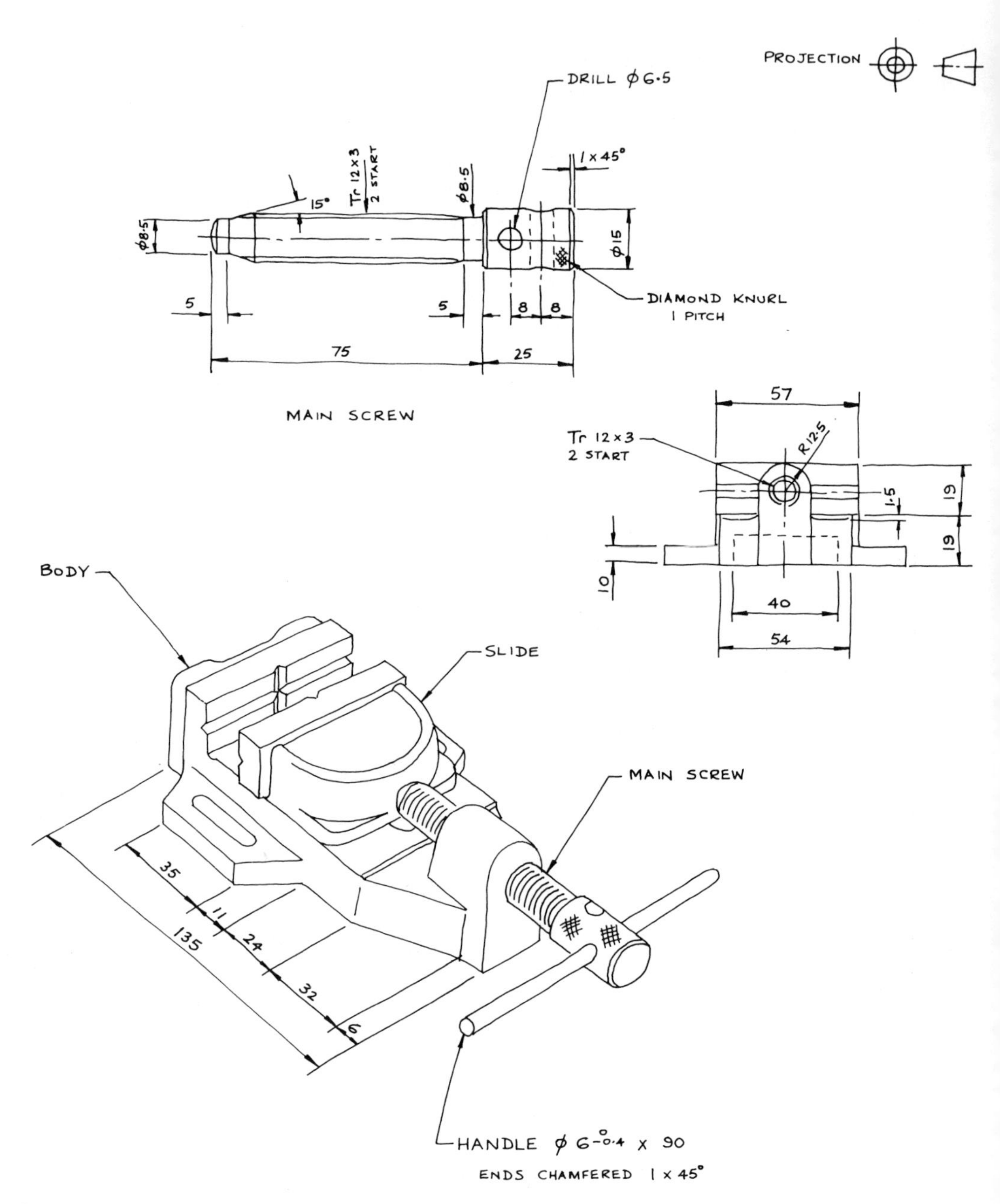

DIMENSIONS IN MM

TOLERANCES UNLESS OTHERWISE STATED
DIMENSIONAL ± 0.4 ANGULAR ± 1°

BODY: BASE SLIDE FACE AND JAW VEE TO BE PARALLEL WITHIN 0.1 TOTAL
JAW FACE AND VERTICAL VEE TO BE SQUARE TO BASE WITHIN 0.1 TOTAL
12.5 SLOT TO BE SQUARE TO JAW WITHIN 0.1 TOTAL

SLIDE: JAW FACE AND VERTICAL VEE TO BE SQUARE TO BASE WITHIN 0.1 TOTAL
JAW VEE TO BE SQUARE TO BASE WITHIN 0.1 TOTAL

Fig. 6.27 Drill press vice.

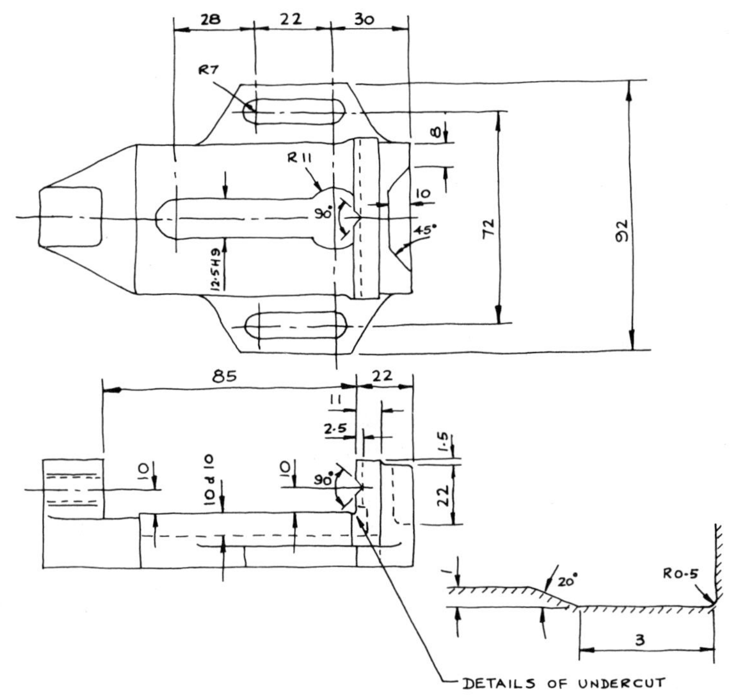
28
22
30
R7
R11
8
10
90°
45°
72
92
12·5H9
85
22
11
2·5
1·5
10
10d10
10
90°
22
1
20°
R0·5
3
DETAILS OF UNDERCUT

BODY

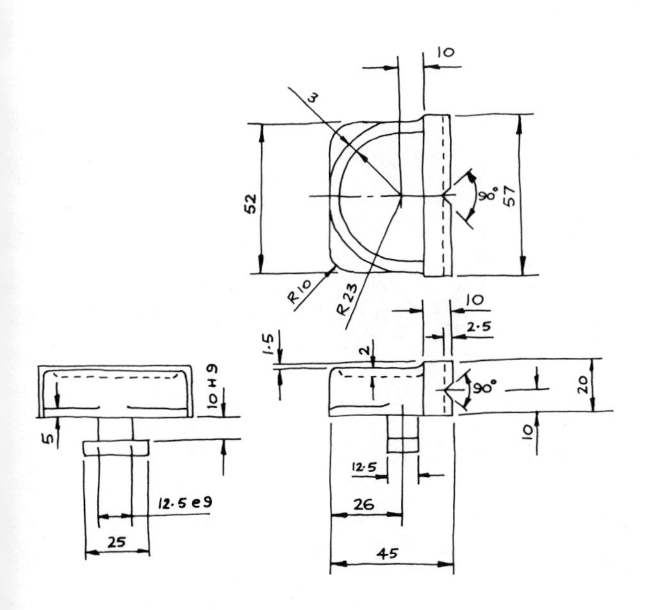
10
3
52
90°
57
R10
R23
10
2·5
1·5
2
90°
20
10H9
5
10
12·5
12·5e9
26
25
45

SLIDE

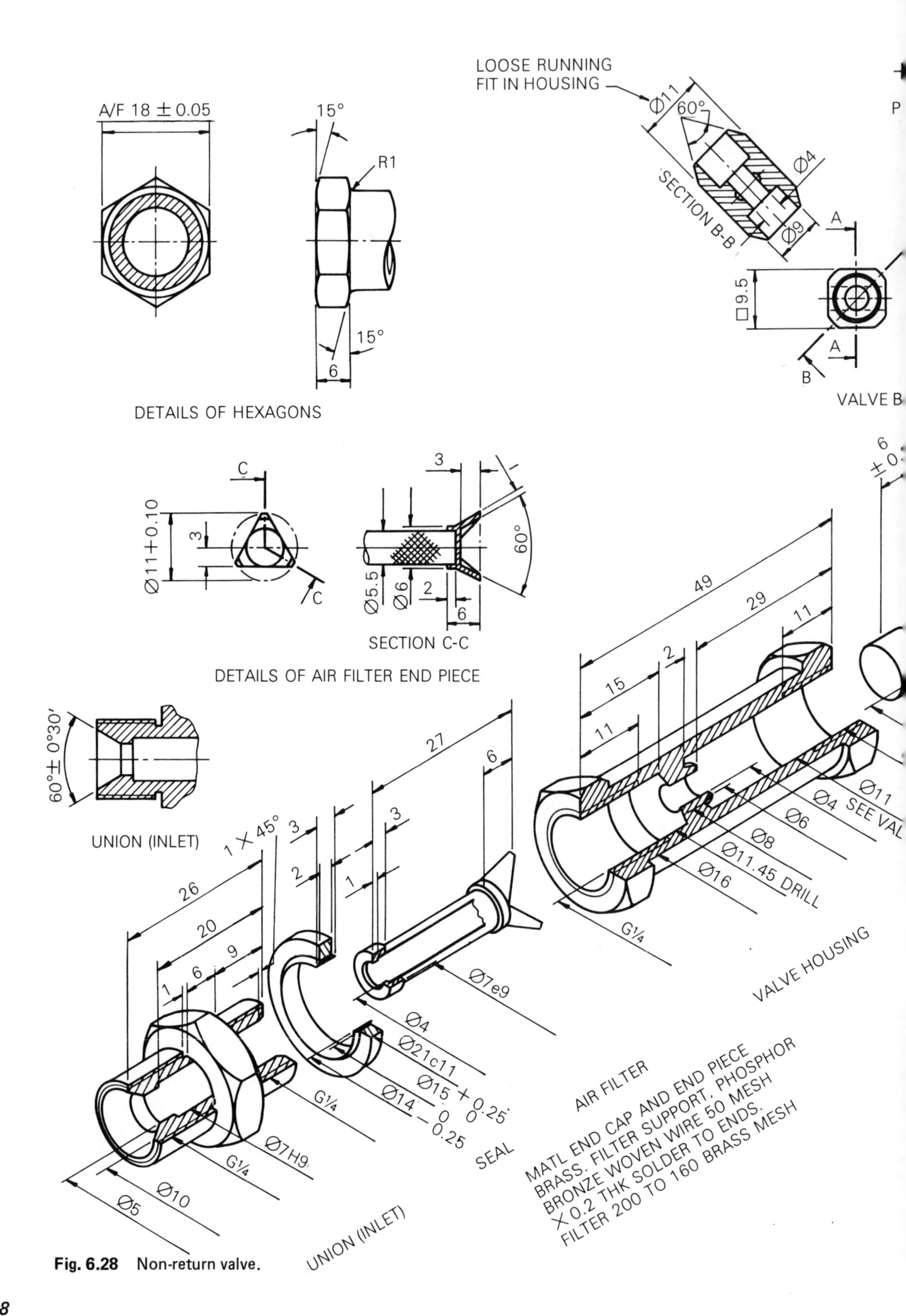

Fig. 6.28 Non-return valve.

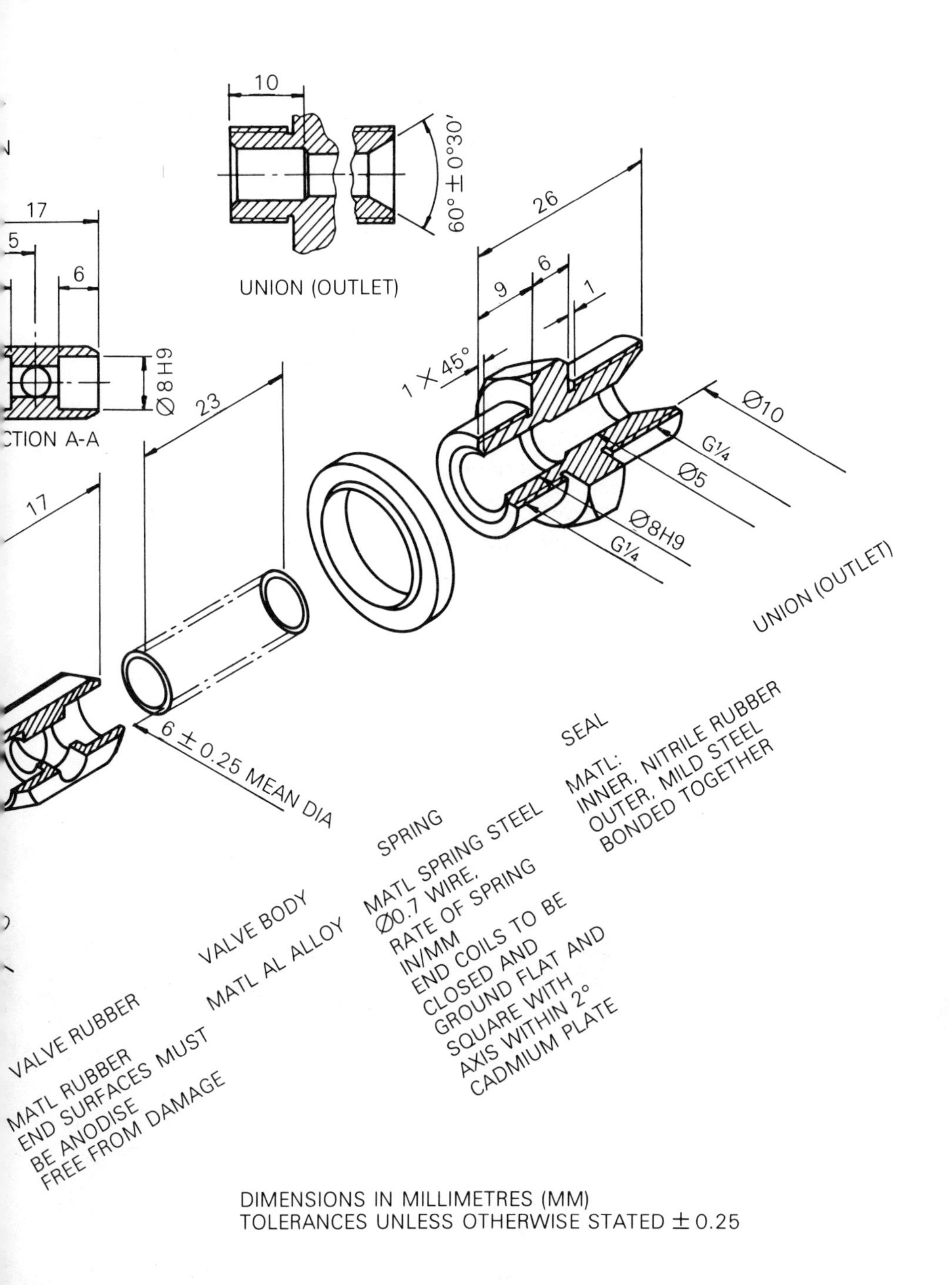

DIMENSIONS IN MILLIMETRES (MM)
TOLERANCES UNLESS OTHERWISE STATED ± 0.25

E THREAD DETAILS:

NOMINAL SIZE	NUMBER OF THREADS PER INCH	PITCH MM	DEPTH OF THREAD MM	MAJOR DIA MM	PITCH DIA MM	MINOR DIA MM
¼	19	1.337	0.856	13.157	12.301	11.445

8. An exploded view of a non-return valve is shown in Fig. 6.28. With the parts correctly assembled, draw, at scale 5:1, a half-sectional assembly drawing. Write a parts list on this drawing and add a title block to British Standard Specification.

9. Produce a fully dimensioned working drawing for each component of the non-return valve in Question 8 to British Standard Specification. Tolerances to be included to BS 308: Part 3: 1972 (ISO). Select suitable material(s) for the unions and housing and state which mechanical property of each material, for all components, makes it appropriate for that use. Add a title block.

10. Visit the College workshop and sketch a lathe tail stock in pictorial view. Dismantle the tail stock and make a sketch of each component in orthographic projection adding all dimensions. Note the type of fit between mating parts.

11. Produce an end-view and half-sectional front elevation (taken through the centre-line of the barrel) of the tail stock in Question 10. Write a parts list on this drawing and add a title block to British Standard Specification.

12. Produce a fully dimensioned working drawing for each component of the tail stock in Question 10 to British Standard Specification. Tolerances to be included to BS 308: Part 3: 1972 (ISO). Select suitable material for each component and add a title block.

Answers

SELF-TEST 1

1. (b)	2. (c)	3. (a)	4. (c)	5. (b)
6. (b)	7. (d)	8. (a)	9. (c), (d)	10. (c)

EXERCISE 1

9. 667 mm, 1.93 seconds

SELF-TEST 2

1. (d)	2. (d)	3. (b), (d)	4. (c)	5. (c)
6. (a)	7. (b)	8. (b)	9. (a)	10. (a)

SELF-TEST 3

1. (c)	2. (b)	3. (a)	4. (a)	5. (d)
6. (b)	7. (c)	8. (b)	9. (a), (d)	10. (a)

EXERCISE 3

1. (a) ISO metric or unified, (b) Whitworth
2. (a) Acme, (b) ISO metric trapezoidal
3. (a) pitch, (b) crest, (c) angle of thread, (d) root, (e) flanks, (f) minor diameter, (g) effective or pitch diameter, (h) major diameter
4. 3
10. (a) metric fine, (b) buttress, (c) trapezoidal, (d) metric coarse, (e) UNC, (f) metric fine

SELF-TEST 4

1. (b)	2. (a)	3. (d)	4. (d)	5. (a)
6. (c)	7. (d)	8. (a), (c)	9. (b), (d)	10. (b)

SELF-TEST 5

1. (a), (b), (d)	2. (b)	3. (c)	4. (d)	5. (c)
6. (b)	7. (c)	8. (a)	9. (a)	10. (d)

EXERCISE 5

1. Clearance fit. Hole: max ∅116.035 min ∅116.000
   Shaft: max ∅115.988 min ∅115.966
4. (a) transition, (b) clearance, (c) transition, (d) interference, (e) interference
5.

	Hole		*Shaft*	
	max	*min*	*max*	*min*
(a)	210.046	210.000	210.033	210.004
(b)	120.054	120.000	119.964	119.929
(c)	2.010	2.000	2.006	2.000
(d)	3.010	3.000	3.020	3.014
(e)	316.057	316.000	316.226	316.190

7. (a) v, (b) (iii), (c) (vi), (d) (ii) (e) (vii), (f) (iv), (g) (i)

SELF-TEST 6

1. (b)	**2.** (d)	**3.** (c)	**4.** (c)	**5.** (a)
6. (a)	**7.** (d)	**8.** (b)	**9.** (c)	**10.** (a), (c)

Index